Contents

Disclaimer

The material in this book is presented after the exercise of care in its preparation, compilation and issue. However, it is presented without any liability whatsoever in its application and use. The contents reflect the personal views of the authors and are not intended to represent those of The Pillsbury Company, Reading Scientific Services Ltd or their affiliates.

Preface

Objective

The food industry is peppered with acronyms and jargon, systems and software solutions many of which claim to assist in the prevention and control of foodborne disease. However, in reality to an outsider they perhaps add to the air of mystery and also could create the suspicion of trickery or deception.

The aim of the Food Industry Briefing Series is to provide a concise, easy to use, quick reference book aimed at busy food industry professionals or students who need to gain a working knowledge. *Food Industry Briefing Series: HACCP* is an introductory level text for readers who are unfamiliar with the subject either because they have never come across it or because they need to be reminded.

Readers who go on to become practitioners in the area of HACCP will need to further their understanding through attendance at symposia, training courses and use of more detailed texts.

Book format

The book is structured such that the reader should be able to skim through in a few hours (perhaps on a train, at an airport, at home in the evening) and arm themselves with the essentials of the topic. In order to achieve this we decided to make it 'non-sequential', i.e. the reader doesn't need to read the whole book from beginning to end in order to grasp what HACCP is about. Instead, we chose an expanding modular format as shown in *Fig. 1*. However, if the reader decides not to read the book from the beginning, but choose certain parts, then any acronyms or specific terms encountered will be explained in the glossary in Appendix B.

Section 1 is the shortest section and it contains many of the questions typically asked by newcomers to the topic – as well as the answers.

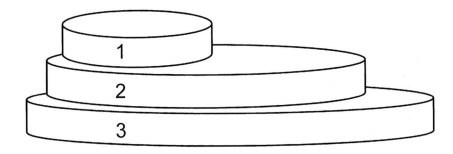

Section 1: Brief introduction to HACCP (who, what, why)
Section 2: HACCP system explained
Section 3: HACCP in practice

Fig. 1 Book layout – expanding modular format.

Section 2 begins to build up in detail and explains the HACCP System in relation to other programmes likely to be in operation in a food business. *Section 3* is the largest section. It looks at how a company would develop a HACCP system, step by step and includes elements of a Case Study which is given in full in Appendix A.

After the three explanatory sections we have written a short Epilogue where current debates in this topic area are discussed along with likely future developments in the field.

The book includes checklists, bullet points, flow charts and schematic diagrams for quick reference. At the start of each section we have provided 'Key Points' summary boxes. These act not only to inform the reader of what the section will cover but also will be a useful way of going back to re-read any particular topic area.

Summary

This book *is*:

- A quick, easy to use reference book.
- Aimed at people who need a working knowledge.
- An introductory level text.

This book *is not*:

HACCP

- The only learning resource material that should be used by those aiming to be practitioners in the field.
- Highly technical.

We hope that we have succeeded in meeting these criteria and that you benefit from reading it.

Sara Mortimore
Carol Wallace
Christos Cassianos

Series Editor's Foreword

All food businesses today operate in complex technical and commercial environments and, if they are to survive, each food business must possess the skills and knowledge required to remain ahead of competitors. Many businesses, and particularly the larger businesses, employ staff who are specialists in the diverse fields that support the food industry's enterprise: from food science and technology, microbiology and engineering to marketing management, logistics and operations management. Yet, though a business may be well served by highly qualified and capable staff, expert in their given field, the needs of the business may not always be fully met. When new skills are required an option is to recruit new staff with the appropriate expertise, but even for the larger business this may not be the preferred option. For the smaller business it is often out of the question. An alternative, and one frequently taken today, is to give staff the development opportunity to learn about the subjects and develop the skills required. In this way food businesses can efficiently and effectively acquire the expertise they need to develop and grow, and to operate within the increasingly demanding frameworks of the established law.

Of the problems faced by staff who desire to learn about new subjects for job development – or who just want to dip into new fields to broaden their knowledge – the choice of study material can be a significant barrier. With so much information available today choosing the right material to study and finding material that can be quickly and easily assimilated can become a complex and off-putting task. In recognition of these issues, the *Food Industry Briefing Series* was devised to assist the food industry in staff development and to provide a useful resource for food industry staff. It is intended primarily for the use of executives, managers and supervisors, but should also find application in academia. Each volume will cover a given topic related to the activities of the food industry and each will aim to provide the essence of the subject matter for ready assimilation either for use in its own right or to create a foundation upon which to layer the concepts contained in more academically demanding texts. The *Food Industry*

Briefing Series is intended to make it easy for the reader to become conversant with, and develop a practical understanding of a particular subject, such that the information and ideas gained can be immediately and confidently applied in any setting, from the factory floor to the boardroom.

This text on the Hazard Analysis Critical Control Point (HACCP) system is the first book in the *Food Industry Briefing Series*. It is written by Sara Mortimore and Carol Wallace who have been able to draw on their considerable experience in food safety management, the implementation of HACCP systems, and the training of HACCP practitioners. The book is clearly written, concise and easy to read. The subject of HACCP is treated in a simple and pragmatic way and, effectively, the reader is taken on a conceptual journey which passes every milestone and important feature of the landscape at a pace which is both comfortable and productive. On completion of the journey the reader should have gained a confident grasp of the theory of HACCP and feel able to take their new-found knowledge into the work place for use in the development and implementation of HACCP systems. When applied to relatively simple food products and their associated manufacturing processes, the information and ideas contained in this book can create the foundation on which new HACCP practitioners will build their confidence and experience, and very soon they should be able to tackle more complex matters. The book should also make a valuable addition to the materials used in staff training and can make an excellent core text for HACCP courses. For those who must know about HACCP, but who have no need to be involved in the implementation of HACCP systems, they cannot go far wrong by reading this book as they will gain knowledge sufficient to converse with even the most expert HACCP practitioner without feeling out of their depth. I am confident that this book will be an asset to the food industry and that it will become thought of as *the* key text on how to do HACCP.

Ralph Early
Series Editor, *Food Industry Briefing Series*
Harper Adams University College
October 2001

Section 1
Introduction to HACCP

Key points

- HACCP is an acronym for the Hazard Analysis and Critical Control Point system.
- It provides structure for objective assessment of 'what can go wrong' and requires controls to be put in place to prevent problems.
- HACCP is a preventative food safety management system.
- It originated as part of the USA manned space programme.
- It is recognised internationally as the most effective way to produce safe food.
- The HACCP principles apply a logical and common sense approach to food control.
- The application of HACCP is possible throughout the food supply chain from primary production (farmers, growers) to the consumer.
- Because it is a step by step approach it is less likely that hazards will be missed. HACCP, therefore, offers increased confidence to the food business and its customers.
- HACCP is cost effective through prevention of waste and incident costs.
- HACCP helps to demonstrate due diligence where required.

Frequently asked questions

By way of introduction to this book and to the subject of HACCP we have attempted to answer some of the most commonly asked questions about HACCP: what it is, how it works, what it looks like, and so on.

1.1 What is HACCP?

HACCP is an acronym used to describe the Hazard Analysis and Critical Control Point system. The HACCP concept is a systematic approach to food safety management based on recognised principles which aim to identify the hazards that are likely to occur at any stage in the food supply chain and put into place controls that will prevent them from happening. HACCP is very logical and covers all stages of food production from the growing stage to the consumer, including all the intermediate processing and distribution activities.

1.2 Where did it come from?

The HACCP concept was originated in the early 1960s by The Pillsbury Company working along with the National Aeronautic and Space Administration (NASA) and the USA Army Laboratories. It was based on the engineering concept of failure, mode and effect analysis (FMEA), which looks at what could potentially go wrong at each stage in an operation and puts effective control mechanisms into place. This was adapted into a microbiological safety system in the early days of the USA manned space programme to ensure the safety of food for the astronauts, in order to minimise the risk of a food poisoning outbreak in space. At that time food safety and quality systems were generally based on end product testing but the limitations of sampling and testing mean that it is difficult to assure food safety. It became clear that there was a need for something different, a practical and preventative approach that would give a high level of food safety assurance – the HACCP system.

Whilst the system was not launched publicly until the 1970s, it has since achieved international acceptance and the HACCP approach towards production of safe food has been recognised by the World Health Organisation (WHO) as being the most effective means of controlling foodborne disease.

1.3 How does it work?

In brief, HACCP is a structured, logical technique applied by following a few straightforward steps:

1. Looking at how the product is made – from start to finish and step by step, identifying possible hazards, deciding at what step in the process they are likely to occur and putting in controls to prevent these hazards from occurring.
2. Deciding which of these controls are absolutely critical to food safety.
3. Setting a limit for safety for the operation of these critical controls.
4. Monitoring these controls to make sure that they do not exceed the safety limit.
5. Identifying the likely corrective action should something go wrong.
6. Documenting the requirements and recording all findings as the products are produced.
7. Ensuring that the system works effectively through regular reviewing of performance and auditing.

These logical steps form the basis of the by now well known seven principles of HACCP which are accepted internationally. They have been published by the Codex Alimentarius Commission (Codex 1993, 1997b) which is the food code established by the Food and Agriculture Organisation (FAO) of the United Nations and the World Health Organisation (WHO) and also by the National Advisory Committee on Microbiological Criteria for Foods (NACMCF 1992, 1997) in the USA. The HACCP principles outline how to establish, implement and maintain a HACCP system. Codex and NACMCF are the two main reference documents and are very similar in their approach.

1.4 What are the seven HACCP principles?

The principles (Codex 1997b) are as follows:

Principle 1: Conduct a hazard analysis.
Principle 2: Determine the critical control points (CCPs).
Principle 3: Establish critical limit(s).
Principle 4: Establish a system to monitor control of the CCPs.
Principle 5: Establish the corrective action to be taken when monitoring indicates that a particular CCP is not under control.
Principle 6: Establish procedures for verification to confirm that the HACCP system is working effectively.
Principle 7: Establish documentation concerning all procedures and records appropriate to these principles and their application.

1.5 Is it difficult to use?

HACCP is often thought of as being complicated, requiring unlimited resources and the expertise associated with large companies. Several specialist skills are, indeed, required in order to use the HACCP principles successfully, but the basic requirement is a detailed knowledge of the product, raw materials and manufacturing processes alongside an understanding of whether any situation which may cause a health risk to the consumer is likely to occur in the product and process under consideration. With both training and education all personnel involved in the application of HACCP should be able to understand and apply its concept. However, for small and medium sized enterprises (SMEs) and less developed businesses the application of the HACCP principles is often found to be more difficult than it first appears. There are a number of reasons for this and research into whether HACCP is appropriate for SMEs is still underway. In the view of the authors, it is not the size of the business that makes it difficult, but the lack of knowledge and capability of the people who work within the business and the poor standard of existing systems such as good hygiene practice. This type of situation can be found in any type of company.

1.6 Why use it?

HACCP is a proven food safety management system that is based on prevention. By identifying where in the process the hazards are likely to occur it is possible to put into place the control measures required. This ensures that food safety is managed effectively and reduces reliance on the traditional methods of inspection and testing.

Inspection and testing have traditionally been the methods used in quality control. Exhaustive inspection would appear to be the ultimate approach towards producing a safe product, at least theoretically. In practice, however, it is not so. Take the example of fruit going down a production line where operatives use visual inspection for physical contamination such as leaves, stones, insects etc. The effectiveness of this technique is reduced by several factors such as:

- Distraction of employees by noise, other activities going on around them, people talking.

- The span of human attention when carrying out tedious activities.
- People's varying powers of observation.

To detect chemical and biological hazards 100% testing is simply not possible because such tests are nearly always destructive. Sampling plans are used instead which are based on:

- The ability to detect the hazard reliably using analytical techniques, which vary in sensitivity, specificity, reliability and reproducibility.
- The ability to trap the hazard in the sample chosen for analysis.

Often random sampling is used and the probability of detecting the hazard is therefore low. Use of statistical sampling techniques will increase the probability of detection but it can never be absolute unless the whole batch is analysed.

1.7 What type of company would use HACCP?

HACCP is applicable throughout the food supply chain from raw material production through processing and distribution to final use by the consumer and can also be applied to non-foods such as primary packaging. *Figure 2* shows in a simple way some of the different stages of food production where food safety is a fundamental issue. If hazards are not controlled at any particular part of this chain, problems could occur or increase later on; it is therefore important that control measures are put into place at each stage of the process, adopting a preventative approach for the entire supply chain.

Primary producers

Primary producers may be farmers raising livestock for the meat industry, fish farmers or harvesters or growers of crops, fruit and vegetables. HACCP use is increasing in this sector but has not been well established histori-cally.

- Example of HACCP use:
  A tomato grower may require to spray with pesticide for controlling presence of a particular pest and may identify pesticide contamina as a possible hazard. The control measure is controlled spraying

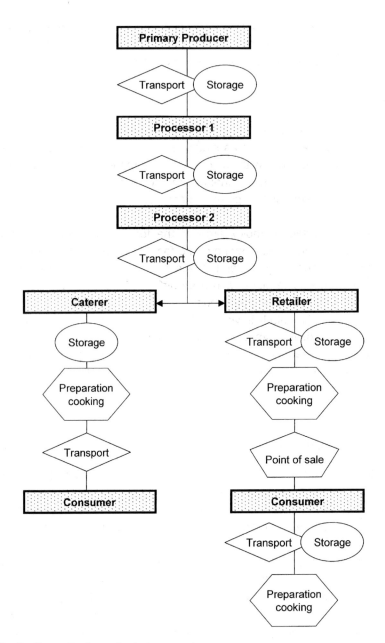

Fig. 2 Example of supply chain.

pesticide. The critical control is monitored by reviewing the signing off of pesticide preparation and application records. The critical limits are the amount and concentration of approved pesticide used together with the length of time before harvest when the pesticide is applied.

It is also important that primary producers are made aware of the impact of their actions further down the supply chain. An issue not identified as a hazard on the farm may have an effect later on in further processing.

Processors

This includes primary and secondary processors of food. Primary processors are operations such as slaughter houses, dairies, sugar and oil refineries etc., who process the raw materials from the farm into a form that can be used further down the chain by the secondary processors. Secondary processors are finished product manufacturers and packers.

This is a particularly complex area of the food supply chain because ingredients used in the final stage of the food manufacturing process may have already been through several stages of primary conversion carried out at different processing plants and even different countries. The potential hazards associated with storage and transportation in such cases must not be overlooked, particularly in view of the variety of climatic conditions and handling involved. It is important that HACCP is used throughout these stages so that hazards can be prevented and any problems that may occur can be traced to their source. It is in this area of the supply chain that HACCP has been most heavily utilised to date and particularly by larger companies.

Caterers/foodservice operators

Catering and foodservice is an area highly prone to food safety incidents because of its very nature, i.e. many operations often happening at once in a restricted area, a vast number of raw materials being handled, short timescales/high pressure to produce and a high turnover of staff.

Many large catering/foodservice chains have used HACCP to
cal areas requiring control. Its use in smaller catering business
somewhat limited and probably driven by regulatory requiren
they exist. Various catering or foodservice versions of the HACCP approach have been developed. One well known example in the UK was the Assured Safe Catering (ASC) approach (DoH 1993). ASC is similar to HACCP in that each step of the process is analysed for hazards and controls are identified; however, it lacks the prioritisation in terms of qualitative risk assessment that a formal HACCP approach would take in identifying the CCPs. As emphasised in the earlier discussion on SMEs, the skill base in many of the small catering establishments is very limited, which can act as a barrier to

use of HACCP. Nevertheless it can be used very successfully in this sector if knowledgeable people are involved.

Retailers

The essential control measures in retail typically include appropriate temperature control and prevention of cross-contamination. HACCP application may be difficult to achieve in smaller shops where both raw and cooked products are sold by the same staff and from the same counter but in using it there is focus on the really critical aspects of the operation, i.e. where controls *must* be in place to minimise the likelihood of a food safety incident occurring. Some retailers process foods on the premises, e.g. butchers and bakers.

The HACCP approach applies to all sectors of the food industry but it is quite often the smaller companies that experience difficulties in implementing HACCP for a number of reasons, including lack of technical expertise and financial considerations (WHO 1999). Overcoming the difficulties is possible and will result in clear benefits in that the business can really target control at the necessary *critical* points.

Consumers

This is a difficult area, as consumers do not necessarily have access to education and training in food safety as does the food industry. There are many similarities between catering and the way that a domestic kitchen operates and studies carried out demonstrate that it is possible to use HACCP techniques to good effect in a domestic kitchen (Griffith 1994).

1.8 Are there any common misconceptions?

A common misconception is that HACCP by itself will ensure that the end product will be of good quality and will meet all legal requirements. The primary aim of HACCP is to control food safety, i.e. to ensure that all food produced is safe for consumption.

● For example:
Contamination of a cooked meat pie with a micro-organism likely to cause illness is a food safety hazard and should be controlled using

HACCP but an over-baked cake is a quality issue, i.e. it may appear darker and be dry in texture. An under-filled bottle of lemonade is a legal matter, i.e. it does not meet the label quantity declaration.

Another misconception is that HACCP is sometimes mistakenly confused with employee 'Health and Safety'. Such statutory requirements are in place in most countries for employers in all industries, not just food, whether manufacturing goods or providing services but this is *not* HACCP. HACCP is purely a management system controlling the safety of a product that will be consumed and is not concerned with the safe working environment of the people involved in its production.

To develop a successful HACCP system it is essential to understand what constitutes a food safety hazard and how to control it. Non-safety issues are managed by other systems that should not be confused with HACCP and the application of its principles.

1.9 How do we know HACCP works?

There are a number of ways by which companies using HACCP will check that it is working. Typically these might be as follows.

Customer complaint numbers

Using the information provided by customers as evidence that the food preparation is not causing problems. Quality complaints can be used as an indicator of all management controls being properly applied, i.e. if there are quality problems then there *may* also be food safety problems.

Auditing

This is the same as auditing any management system except that the documents prepared using HACCP principles can be assessed for both completeness and compliance.

Test results

Routine and specifically planned tests may be used to verify HACCP effectiveness. Records should be reviewed to make sure that all such tests have been carried out properly and that the results were within specification.

1.10 What actually gets implemented in the workplace?

A HACCP system is summarised in a document known as a HACCP plan. This is simply a document or folder and it contains all the information related to the critical control points (CCPs) – together with the operating standards or critical limits. It also documents who is responsible for the monitoring of the CCPs and at what frequency, what corrective action should be taken if something goes wrong, the hazard that is being controlled, and it often includes a process flow diagram or stepwise drawing of each step in the process. It is actually the CCPs that are implemented in the workplace through the monitoring and corrective action activities.

1.11 How does a HACCP plan get written?

A HACCP plan is the output of the HACCP study. This is the application of the first five of the Codex HACCP principles, i.e. the raw materials and the processes used are evaluated to see what foodborne hazards may be a concern and the appropriate controls are identified.

Before the HACCP study commences a certain amount of planning and preparation will occur – in fact this is important at every stage. Once the HACCP study is completed and implemented it needs to be kept up to date, i.e. verified as being correct and maintained to keep current with the product and process as they undergo changes. In simple terms, the whole process of actually using the HACCP principles could be broken down into four key stages as shown in *Fig. 3*.

1.12 Who carries out the HACCP study?

This is done by the HACCP team. Normally a multi-disciplinary team of about four to six people are trained in a HACCP approach. In a large business the functions included may be quality assurance, manufacturing, engineering, research and development, microbiology and supplier quality assurance. In a smaller business there may only be one or two people available. What is

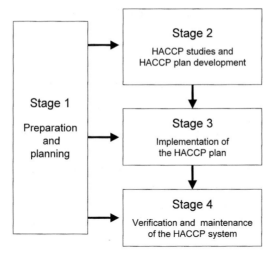

Fig. 3   The four key stages of HACCP (adapted from Mortimore and Wallace 1998).

important is that the people on the team have an in-depth knowledge of how the product is made. They also need technical know-how, particularly for the hazard analysis, but if this is not available within the business then it can often be bought in on a consultancy basis.

1.13 What is the regulatory position of HACCP?

Governments and enforcement authorities are increasingly recognising HACCP as the most effective means of managing food safety.

The European Community Directive 93/43 EC (1993) on the hygiene of foodstuffs states that 'food business operators shall identify any step in their activities critical to ensuring food safety and ensure that adequate safety procedures are identified, implemented, maintained and reviewed'. This is effectively a recommendation to adopt the HACCP approach towards food safety for all food businesses.

In the UK the defence of due diligence contained within the Food Safety Act 1990 requires that the person proves that he took 'all reasonable pre-cautions and exercised all due diligence to avoid the commission of the offence by himself or by a person under his control'. In the event of litigation this would give a defendant a strong case if HACCP was in place and working – and a very weak case if it was not!

In New Zealand the Ministry of Agriculture is in the process of making HACCP mandatory for all food products. Previously a voluntary approach was taken.

In the USA the HACCP techniques were used to identify the controls specified in the Low Acid Canned Food Regulations as early as 1973 (USDA 1973). More recently the USA Department of Agriculture has decreed that HACCP programmes are required for all meat and poultry processing facilities (USDA 1995). HACCP techniques are also required by law in the area of seafood inspection and processing (FSIS 1996). Other food areas will follow. The trend indicates that HACCP could eventually become mandatory in the USA, not only for all USA food processing facilities but also for processors from anywhere in the world exporting to the USA.

While many countries are in the process of re-evaluating and developing their food safety policies, use of the Codex HACCP principles as the international standard means that the HACCP systems implemented by trading partners are based on the same principles. At the time of writing we have some way still to go before equivalency in interpretation and implementation is agreed but the General Agreement of Tariffs and Trades (GATT) Uruguay Round and the establishment of the World Trade Organisation (WTO) in January 1995 has meant that mutual agreement of the standards of each trading partner's country and/or the equivalence of food safety system must occur before trade can proceed.

In summary, it is clear that international legislation is moving more and more towards making HACCP a mandatory requirement in the food industry. This will lead to greater regulatory assessment of HACCP systems as governments take up their responsibilities with regards to confirming that the business operators are properly complying with requirements (WHO 1998).

1.14 Are there other driving forces for the use of HACCP?

Customers and consumers

While the end consumer may not know what HACCP means, manufacturers, retailers and caterers increasingly expect that their suppliers develop and implement HACCP plans. Any inspection carried out on production premises nowadays includes an assessment of the competence of the management. An effective HACCP system is essential in demonstrating that the food business operator understands and is managing food safety hazards.

Media pressure

As consumers become more aware of food safety they are encouraged to go to the press, often lured by the publicity and potential cash rewards.

The issues identified may not always be real in terms of food safety but can cause severe brand damage and food suppliers need to be able to answer all claims made against them. In addition to providing an effective food safety protection programme, fully documented evidence in the form of efficiently maintained HACCP records may be essential in counteracting such claims and ensuring that the company stays in business.

1.15 What does it cost?

There is no fixed price that can be put on a HACCP system. Whichever way one looks at it, it is going to be variable depending on what is already in place and the complexity of the process.

At the planning stages of a HACCP system it may appear that a lot of expense is required, but on reflection much of this is not HACCP cost.

For example:
HACCP may identify the need to improve practices within the operation and training of staff, but these requirements existed anyway; HACCP has simply highlighted them.

The true cost of HACCP will include:

- Time during HACCP training.
- HACCP training – external courses or hiring of a trainer.
- Administrative support.
- Additional temporary resources (e.g. technical, secretarial).
- Cost of validation.
- Time cost for review/audit.
- Equipment – if identified as a need.
- Verification activities.

Some of these costs, however, may well be offset by the savings resulting from the application of HACCP such as:

- Reduction of on-line product testing in terms of product used and human resource.
- Possible reduction in analysis costs, both internal and external, if the HACCP study identifies alternative measures, e.g. certification of ingredients.
- Earlier release of finished product, thus reducing stock holdings.
- Hazards have major implications for food manufacturers and retailers and the financial cost that ensues when control is lost can run up to a fortune in lost sales, court costs and compensation, plus loss of confidence and its effects.

Additional costs need to be budgeted at the planning stage to ensure that the project is fully financed to its completion and subsequent maintenance.

1.16 Is there anything more that I should know?

HACCP is not an exact science. It is a tool, a way of thinking – where decisions taken must be based on sound science but even after 40 years its use and interpretation of the principles continue to be debated across international boundaries. This is worth bearing in mind when going into more detailed sections of the book.

Section 2
The HACCP System Explained

Key points

- HACCP system development requires skills, activities and conditions that are not unique to HACCP alone.
- Normal management practices such as project management and leadership skills are a key success factor.
- Prerequisite good hygiene practice programmes support HACCP but also control quality or 'wholesomeness'.
- Quality management systems will provide an effective framework within which to implement HACCP.
- Management commitment is essential if HACCP is to be taken seriously.
- The HACCP team must have a broad variety of expertise and may include use of external food safety experts.
- The structure of the HACCP plan must be decided based on the nature and complexity of the products and processes.
- It is important that the HACCP project is planned carefully in advance.

This section is divided into three parts:

2.1 HACCP system overview – how does it all fit together?
2.2 HACCP in the context of other management activities – what is HACCP and what is it not?
2.3 What is involved in getting started – i.e. the preparation and planning stage.

2.1 HACCP system overview – How does it all fit together?

In the previous section a number of key terms were introduced – part of the HACCP jargon. To recap:

HACCP principles – there are seven principles published by Codex Alimentarius (1997b) and NACMCF (USA) (1997). They are:

Principle 1: Conduct a hazard analysis.
Principle 2: Determine the critical control points (CCPs).
Principle 3: Establish critical limit(s).
Principle 4: Establish a system to monitor control of the CCP.
Principle 5: Establish the corrective action to be taken when monitoring indicates that a particular CCP is not under control.
Principle 6: Establish procedures for verification to confirm that the HACCP system is working effectively.
Principle 7: Establish documentation concerning all procedures and records appropriate to these principles and their application.

HACCP study – this is where the HACCP principles are applied to a segment or whole of the food chain under consideration.
HACCP plan – the output of the HACCP study. This is a document prepared in accordance with the principles of HACCP to ensure control of hazards which are significant for food safety.
HACCP team – the people who carry out the HACCP study, usually four to six people making up a multi-disciplinary team. Typically quality assurance, engineering and production personnel will be on the team.
HACCP system – the HACCP system is what you get once the HACCP plan has been implemented in the workplace.
HACCP documentation and records – used as evidence that the system is in place and working.

An effective HACCP system needs to be planned, studied, implemented, verified and maintained in a logical systematic way. *Figure 4* shows how the various elements fit together.

Preparation and planning are very important and can be a key element in ensuring that the process of setting up a HACCP system is as painless as possible. Later in this section we will look at this activity in more detail.

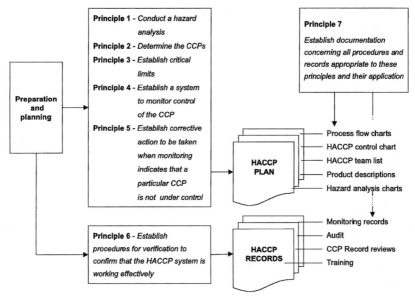

Fig. 4 HACCP system overview.

Before that let's look very briefly at what the application of the HACCP principles involves.

The HACCP study itself is essentially made up of the first five of the seven principles. A HACCP team will start by applying the first principle and in order to do this they will map out the food process step by step. This is the HACCP study:

● Example: Boiling an egg
The process of boiling an egg can be broken down and documented on a process flow chart as shown in *Fig. 5*.

At each of the six process steps identified the HACCP team will assess whether there are any food safety hazards of concern (Principle 1). So at step 1 they may identify *Salmonella* as a potential problem.

They will then consider whether the step of removing the egg from the refrigerator is a critical step (Principle 2) with respect to the control of the hazard in question, i.e. *Salmonella*. They are likely to conclude that it is not, given that the egg is going to be boiled in water later on. At step 5 (where the egg is boiled), as they apply Principle 2 they are likely to conclude that the act of boiling is critical (a critical control point) for control of *Salmonella* – it is a micro-organism which is easily destroyed by thorough cooking.

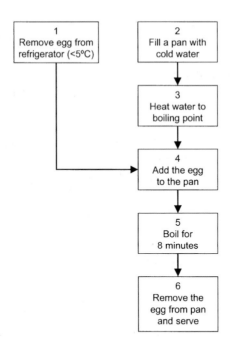

Fig. 5 Boiling an egg – process flow chart.

They will then consider how long the egg must be boiled (the critical limit) to ensure that any *Salmonella* is destroyed (Principle 3). In manufacturing facilities this is usually determined through various tests and measurements, e.g. to assess the centre temperature of the egg over the course of the cooking time in relation to known information on thermal destruction of *Salmonella,* and therefore that the boiling time chosen is sufficient to reach the required centre temperature.

Having proven the relationship between time and temperature, the HACCP team has to decide how often to monitor the boiling time and the water temperature (Principle 4) and what to do if the requirements are not achieved (Principle 5).

All this information will be documented on a form often known as a HACCP control chart or worksheet that goes into the HACCP plan. Other pieces of information may also be recorded and retained within HACCP plans, e.g. details of who was on the HACCP team, a description of the product concerned and the notes taken during the hazard analysis. There are no hard and fast rules on what additional information to keep.

Once the study is complete the team will need to carry out validation activities to confirm that all elements of the HACCP plan will be effective.

⬤ Example: Boiling an egg
Validation: Confirmation that boiling an egg for 8 minutes will destroy *Salmonella.*

Verification activities (Principle 6) typically include tests, random sampling and analysis, reviews of monitoring records and audits – all designed to determine whether the HACCP system is working effectively once implemented in the operation.

⬤ Example: Boiling an egg
Verification will include a review of the monitoring records which show that the egg was boiled for 8 minutes.

It is the verification activities that principally lead to the compilation of a number of documents and records which is another (Principle 7) requirement, though the HACCP plan itself is obviously a key document (see Fig. 4).

The HACCP principles are logical and it is an easy to understand concept. However, even though the example given (boiling an egg) is very simple it is clear that some technical knowledge is required, i.e. that *Salmonella* is associated with raw eggs and that boiling for 8 minutes will destroy it. Many people with basic hygiene knowledge (or cooking ability in this case) might know some of this but consider the situation if the product was sushi or a pizza with a range of toppings – it is not so easy. There is also a whole range of other skills and activities that are needed when HACCP is put into business for real.

2.2 HACCP in the context of other management systems – What is HACCP and what is it not?

This comes as a surprise to some people but HACCP by itself will not guarantee safe food. In order to be effectively developed and implemented HACCP needs the support provided by other management systems operating within the business. These can be categorised into four groups as shown in *Fig. 6.*

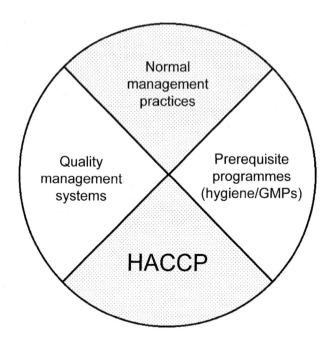

Fig. 6 HACCP in the context of other management systems.

Normal management practices are the day-to-day 'business' skills which, when thinking about HACCP as a project, will be invaluable in making sure that the work involved is carried out efficiently and effectively. Prerequisite programmes are basically the good manufacturing practices (GMPs) or good hygiene practices (GHPs) that any reputable food business operator would be adhering to in order to ensure that safe, wholesome food is provided to the consumer. Quality management systems act as a framework within which any process activity can be managed, including HACCP.

2.2.1 Normal management practices

There is a vast array of skills and activities which contribute to managing a business, many of which may not be formally recognised, particularly in a small business. Some of the skills, however, are needed during HACCP development and implementation. A few of these are considered below.

Management commitment

Management commitment is needed for any project undertaken by a business and should be the driving force if the project is to be fully successful. Real commitment for HACCP will only be achieved if the management team understands fully what HACCP is all about – the reason for using it, the expected benefits, what is involved, how long it might take, what costs would be incurred, any likely impact on other aspects of the business (e.g. improvements in GMP) and so on. It will also be helpful to confirm what has contributed to the decision to use HACCP – legislation, customer demand or a desire for self-improvement.

Leadership

The HACCP team will lead the process of developing and implementing a HACCP system. Leadership skills will not only be relevant within the senior management of the business but also within the HACCP team itself and will need to be further developed if not already in place. HACCP implementation is a good opportunity for motivating the workforce by reinforcing hygiene standards and making them proud of being part of the organisation. This can be achieved through strong leadership.

Project management

As will be discussed later in the preparation and planning stage, putting a HACCP system into a business is a major project and needs to be managed as such. The work involved can be considerable, particularly in a complex business.

Process mapping

Process mapping is used widely in business improvement programmes as it is a systematic way of getting to understand a given process. It can be used for any process, e.g. activities as diverse as 'customer order processing' through to employee medical surveillance. Process mapping is of course one of the first activities undertaken by a HACCP team when they begin the hazard analysis step by constructing a process flow diagram.

Data analysis and administration

Data analysis and administration are used in a number of business situations (e.g. accounting, labour and overheads management) and also in HACCP verification where CCP records are reviewed and customer complaints analysed. Simple statistical calculations can be helpful in identifying developing trends.

Problem solving

Problem solving is a very useful skill set, frequently used in HACCP as in many other projects. It may involve a variety of techniques such as identifying the 'desired future state' and then analysing the drivers and restrainers that either prevent or expedite you getting there. It could also encompass 'cause and effect' analysis – often run as a facilitated brainstorming session with all interested contribution parties.

Auditing

Again a key activity in HACCP verification, auditing is used to measure compliance with the documented HACCP plan, but is also a traditional management activity, for example, in accounting as well as in quality management.

Team skills

It is useful to build the HACCP team so that they enjoy a more efficient relationship based on openness and trust, all working towards the same goal. Again, these skills are needed in a number of other business teams and are not specific to HACCP alone.

Record keeping and documentation

Record keeping and documentation, while a HACCP principle, is actually an activity which is not exclusive to the administration of HACCP. There may be other skills and activities which are needed if a really sound HACCP system is to be developed and implemented, and which will provide a backbone for the company's food safety programmes. We have highlighted a few of the more obvious ones.

2.2.2 Prerequisite programmes

The World Health Organisation (WHO 1999) defines prerequisite programmes as 'Practices and conditions needed prior to and during the implementation of HACCP and which are essential for food safety'. There is nothing new about prerequisite programmes but the term itself is fairly new. In simple terms prerequisite programmes are those practices which many would class as good manufacturing practices (GMPs) or good hygiene practices (GHPs).

The requirement to have an environment which is operating to good standards of hygiene and housekeeping is clearly fundamental to the day to day management of food safety and wholesomeness. *Figure 7* shows some of the prerequisite elements that provide essential support to an effective HACCP system. Prerequisite programmes control the general factory or kitchen 'good housekeeping' issues rather than specific process hazards which are managed through HACCP, and much of this is common sense.

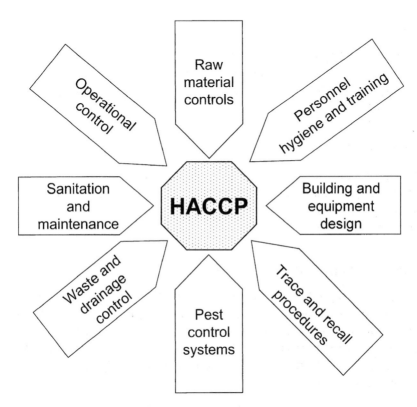

Fig. 7   Hygiene prerequisites for HACCP (based on Codex Alimentarius Commission Food Hygiene Basic Texts (Codex 1997a)).

Attention must be given to the location of a food handling facility, also good hygienic design and construction and the provision of hygienic facilities such that hygiene is given a high priority, enabling the hazards to be effectively prevented and controlled. This will include:

- The fabric and layout of the building (i.e. walls, floors, ceilings).
- Provision of handwash facilities and toilets.

- Pest proofing.
- Hygienic design of equipment.
- Water supply.
- Food and waste container design and availability.
- Drainage and waste disposal.
- Air quality and ventilation.
- Adequate lighting.
- Temperature control and storage facilities.

As well as ensuring good facility design it is important to control hygiene in the other elements of operations such as formulation design, raw materials, processing and packaging. Key areas will include:

- Time and temperature control.
- Cross-contamination control (microbiological, chemical and physical).
- Control of incoming materials.
- Packaging materials control.
- Water (all areas, i.e. as an ingredient and for sanitation).
- Documentation and record keeping.

Waste removal and drains can be ready sources of contamination if not managed. Waste should be covered and regularly removed from the facility so as not to attract pests. Drains should be kept clean and the direction of flow should be such that cross-contamination from 'dirty' to 'clean' areas does not occur.

In the USA sanitation programmes (cleaning and disinfection) are a legislated prerequisite in the meat and poultry industries, where HACCP is regulated, and are known as sanitary standard operating procedures (SSOPs) – another acronym which is often heard in HACCP circles. This is an important aspect of a hygiene control programme in any food business and an area that can often be overlooked given that sanitation frequently occurs on a night shift and, if it is not given a high profile, often receives a low priority. There are many examples of food safety (and spoilage) recalls that are a result of poor sanitation.

Effective pest control programmes are essential in the food business and should include both preventative and corrective action programmes for insects, rodents, birds and other appropriate pests (e.g. snakes). Pests can act as carriers of microbiological contamination (which is the primary concern) as well as being a source of physical contamination which is aesthetically undesirable either through presence of particulates or through being the source of pack damage.

Personal hygiene and hygiene education and training are very important in that the hygiene standards of a business will only ever be as good as the hygienic behaviour of the people who work within it. Inadequate hygiene is a known major source of cross-contamination, either directly (e.g. through sneezing over food, spitting or smoking) or indirectly (e.g. through lack of hand washing after using the toilet). It is essential that the company is able to rely on a consistent hygienic behaviour from the workforce. Jewellery and other personal effects should not be allowed in the workplace as they can act as an indirect source of microbiological contamination as well as potentially being a physical contaminant. Any visitors to a food handling area should also comply with hygiene rules regarding handwashing, absence of jewellery, the wearing of clean protective clothing, covering of hair and so on.

Raw material control is another essential aspect of a prerequisite pro-gramme. The approval of suppliers of critical raw materials is often done through an audit of the supplier's premises, particularly in a large company, and raw material specifications will document all likely hazards together with control procedures and monitoring checks that are carried out by the supplier. Certificates of analysis are often provided by suppliers as evi-dence that their own HACCP systems are working. These should be sup-plied by reputable laboratories if results are to be relied upon. For SMEs reliance on purchase from 'reputable' suppliers is often the best option for ensuring that wholesome raw materials are used.

During the distribution of the food it is important to ensure that it is pro-tected against contamination and that where temperature control is important (i.e. chilled and frozen food) this is managed such that the growth of micro-organisms or production of toxins is prevented. If control is lost it is important to ensure that recall procedures are in place to ensure that any ensuing unsafe product incident is managed effectively and product could be recalled from the marketplace if necessary.

2.2.3 Quality management systems for effective operation and process control

Quality assurance systems, such as those modelled on the International Organisation for Standardisation's ISO 9000 quality system standards, aim at primarily ensuring customer requirements are met consistently. Both HACCP and such quality management systems aim at the prevention of non-conformity, placing emphasis on effective corrective action and getting it right first time.

Whilst a quality management system is not a 'prerequisite programme' in terms of good hygiene practice, it is often used to manage the prerequisite programme and HACCP systems so that any element of the operation can be effectively controlled (*Fig. 8*).

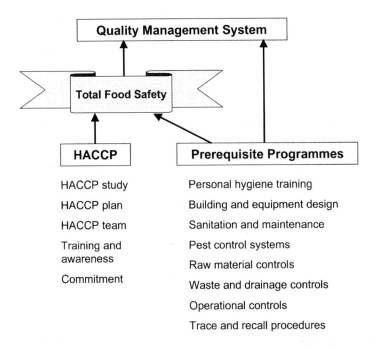

Fig. 8 Quality management programme (from Mortimore (2001)).

There are many controlling steps in any process; some will be critical for food safety but others will control the quality and legal attributes of the product. These are sometimes known as process or legal control points or simply control points (CPs). To recap: critical control points (CCPs) are the stages in a processing operation where the food safety hazards must be controlled. CCPs are essential for product safety as they are the points where control is affected. The CCP itself does not implement control. It is the action that is taken at the CCP that controls the hazard.

Some companies often implement extra control points in their processes in order to protect the process and process equipment or to relieve pressure from the CCP by reducing the hazard. These points are usually CPs and they should not be confused with the genuine CCPs where the hazards *must* be controlled.

This simple relationship between CCPs and CPs must be understood if HACCP is to be used to its best effect. It is important that CCPs are identified as the points which are truly critical to product safety.

Figure 9 serves to illustrate how both CCPs and CPs may be documented and managed within the framework of a quality management system such as ISO 9000.

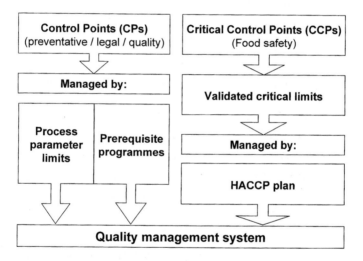

Fig. 9 Control point differentiation (adapted from Mortimore in Wallace and Williams (2001)).

While normal management practices, prerequisite programmes and quality management systems assist in managing both food safety and quality, before considering the development of a HACCP programme it is important that their current status is evaluated in order to determine:

1. What skills, activities and conditions are required to combine with HACCP to enable effective food control?
2. What is actually already available?

The answers to these two questions will highlight the gaps that need to be filled and this evaluation is often called a 'gap analysis'.

When identifying the deficiencies it is important to be clear about what forms part of the HACCP system and what is, or should be, in place as a foundation or support to HACCP implementation. It can be helpful to communicate this relationship to employees early on in the programme.

● For example:
Personnel hygiene practices such as handwashing facilities and suitable protective clothing are of utmost importance in helping to protect products in any food handling operation.
Calibration of equipment which is a requirement of ISO 9000 is also a key requirement when it comes to ensuring that equipment used to control CCPs is working properly.

Understanding of the relationship between these programmes and HACCP is likely to lead to a more structured and systematic approach. It will also aid understanding of what HACCP is really designed to do – identify and control hazards through the management of critical control points, whilst providing a clear understanding of where additional control points can be effectively set up.

2.3 What is involved in getting started – The preparation and planning stage

Proper preparation and planning (*Fig. 10*) are fundamentally important to developing a successful HACCP system. It is essential at the earliest stage of setting up a HACCP system that:

- Senior management commitment is assured.
- The appropriate people are identified and trained.
- The prerequisite support systems already in place are established and what needs to be further developed is planned for.
- The most appropriate structure for the HACCP system is selected after careful consideration.
- The entire project for the development and implementation of the HACCP plan is planned.

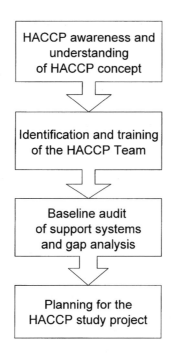

Fig. 10 Key stage 1: preparation and planning (adapted from Mortimore and Wallace (1998)).

Management commitment and training

One of the first preparation activities is to gain an overall awareness and understanding of what is involved in using HACCP. In order to do this properly it is essential to gain commitment at senior level and ensure proper allocation of resources. Real commitment can only be achieved if there is full understanding of HACCP and its benefits and what is involved and required at senior management level. This can be achieved through reading specialist publications and attending a HACCP briefing session undertaken by an expert within the organisation, if available, or through an external consultant, if not. Visiting other companies who have already implemented a system is invaluable in gaining an understanding of what the outcome of the project will look like, i.e. the HACCP system. It is an ideal way of seeing various styles if this is possible, talking to people who have done it, about what worked, what didn't and what they might do differently if they were starting again. This will be useful knowledge at the start of the project.

It is highly desirable that the HACCP study is not carried out by one person alone, though in a SME there may not be a choice. Ideally HACCP development should be carried out by a multi-disciplinary team. Selection of team members should be based on knowledge of raw materials, products, processes and hazards. Ideally the core HACCP team should consist of people having knowledge and expertise in the following areas:

- Quality assurance/technical – providing expert advice in handling microbiological, chemical and physical hazards, an understanding of the risks, and knowledge of the measures needed to control the hazards.
- Operations or production – people who have responsibility and working knowledge of the operational activities needed to produce the product.
- Engineering – to provide a working knowledge of process equipment and environment with respect to hygienic design and process capability.

Additional expertise may be needed depending on the nature and complexity of the product. In a larger company the following areas should be considered:

- Supplier quality assurance – essential in providing details of supplier activities and assessing risks associated with raw materials, particularly where high-risk materials are involved.
- Research and development – particularly where new products are constantly being developed.
- Distribution – especially where temperature control is essential to product safety.
- Purchasing – could be useful where factored goods are purchased and re-sold.
- Microbiologist – if a company has its own microbiologists their expertise will most likely be needed in the HACCP team. Smaller companies who do not have this option should consider an external source of such expertise.
- Toxicologist – may be needed where chemical hazards and their monitoring and control are a potential issue. A toxicologist could be located in a food research association, consulting analytical laboratory or university.
- Statistical process control (SPC) – an external expert may be needed when setting up sampling plans or a detailed assessment of process control data.
- External HACCP 'experts' – may be appropriate initially to guide the HACCP team through the selection of team members and initial studies.

A HACCP team needs to be small enough for effective communication but large enough to enable specific tasks to be delegated. This is why a range of four to six is usually quoted as the ideal number of HACCP team members.

The development, implementation and maintenance of a HACCP system requires some skills and activities that are unique to HACCP and some that are not. *Figure 11* emphasises the skills that are already considered as being normal management practices and those that are unique to the use of the HACCP principles.

One member of the HACCP team should be selected for their leadership skills and appointed as the team leader. The team leader will be responsible for ensuring that:

- The team members have the necessary knowledge and expertise through training and development.
- All tasks relating to the development of HACCP are organised adequately.
- Time is used effectively and also is made available for reviewing progress on an ongoing basis.
- All skills, resources, knowledge and information needed are identified and sourced either from within the company or through external support.
- Documents and records are maintained efficiently.

It is very useful to appoint a 'scribe' or secretary to the HACCP team. It is a key role and requires someone with good attention to detail. Since each part of the HACCP study builds on the previous part and may be completed at different times, accurate record keeping is essential.

Once selected, this team must be prepared with detailed training in the principles of HACCP and with additional training and understanding of the management skills and topics which underlie the application of these principles.

Training of the HACCP team is the single most important element in setting up a HACCP system and it is important it is done properly. External training courses from reputable training providers or regulatory authorities are a good option but often only provide an introduction, so the HACCP team should not be expected to be experts after a two-day course.

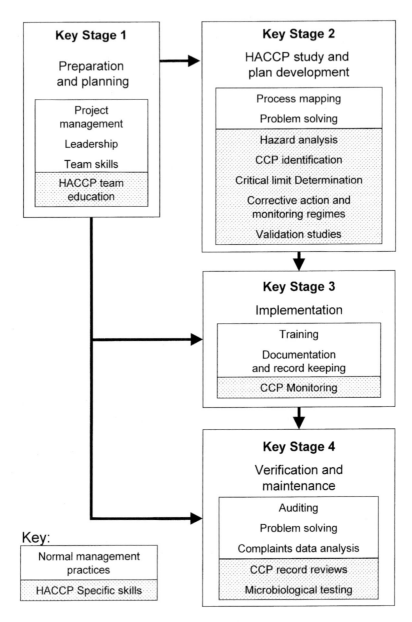

Fig. 11 HACCP system development – skills requirements.

Baseline audit

It is a good practice to conduct a thorough baseline audit of the current systems within the business at the start of the project in order to fully appreciate the scope of what might be involved during the development and implementation of HACCP. Not all companies will do this but the ones who do are usually self-driven, i.e. they want to see real business benefits and do as good a job as possible. Companies who are putting in a HACCP system solely because of legal or customer requirements may want to do the bare minimum to gain compliance, i.e. to produce a HACCP plan which satisfies the regulator or customer.

The baseline audit is a good opportunity to evaluate a number of elements:

- The status of the prerequisite programmes as described earlier.
- The skills base within the organisation and to begin talking about the steps needed to fill the gaps.
- The controls and procedures already in place and which will now be incorporated into the HACCP plan.

Planning the HACCP project

Having gained commitment from senior management and understood the desired outcome of adopting the HACCP approach and trained the HACCP team, the next task is to begin planning to get there.

In planning the HACCP project, the HACCP team leader will need to ensure that the team has a complete understanding of the project vision and knows clearly where they are starting from and what the end result will look like in the factory or catering facility. This is most easily accomplished if the team is appointed early and participates in the initial awareness activities as well as the baseline audit and gap analysis so that they fully appreciate the current capabilities and the size of the task.

One of the key issues to be decided early on is the structure of the HACCP system. This will depend on the complexity of the operation and the types of processes being carried out. There are three basic approaches, as follows.

Linear HACCP plans

In this approach the HACCP principles are applied to each product or process on an individual basis starting with the raw materials coming in and ending with the finished product. This approach works best in simple operations where there are few product types and a small number of processes

● For example:
A bakery producing one type of bread.

Modular HACCP plans

This approach works best where there are several basic processes used to produce a number of products.

● For example:
A factory producing several types of pizza and other dough products where several basic processes are in place and each product will involve a combination of operations.

The HACCP principles are applied separately to each basic process (or module) and these modules are finally combined to make up the complete HACCP system, as shown in *Fig. 12*. In this example the business will have seven separate HACCP plans, i.e. one for each module. Care should be

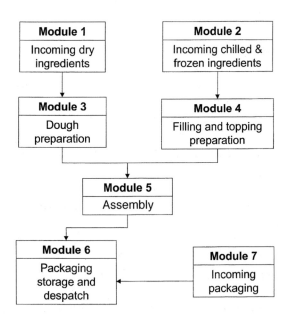

Fig. 12 Modular HACCP plan example – pizza and dough products modules.

taken to define the specific start and end points of each module to ensure that no hazards are missed out.

Generic HACCP plans

This approach is used where similar operations are carried out at different locations in the manufacture or handling of similar products.

● For example:
  Primary meat processing carried out at several processing sites using the same basic methods.
  A fast food restaurant chain using the same ingredients and process steps.

Generic HACCP plans can be either linear or modular and can make a helpful starting point. An effective HACCP system can be built around them by tailoring the generic HACCP plan to the operating requirements, but there are limitations because no two operations are exactly the same and there is a danger that hazards may be overlooked if an 'off the shelf' HACCP plan is used without modification.

Increasingly generic HACCP plans are being developed for sectors of the industry which do not have the ability to develop their own. For example, in the UK, a generic HACCP plan template was developed for the retail butcher trade (MLC 1998) and currently the preparation of a generic HACCP plan for small cheese producers is underway.

These generic plans are arguably better than not having anything at all and could work well if:

1. Mature prerequisite hygiene programmes are in place appropriate to the industry, and
2. They are adapted to fit the local needs – perhaps with the help of the regulatory enforcement authorities, though this activity is not normally regarded as being part of their role (WHO, 1998).

Once the team has agreed the structure of the HACCP system, the HACCP team leader can more properly estimate the resources and time required to complete the task. This will help to ensure that the programme stays on track and that problems are discussed as they are identified. A Gantt chart (see *Fig. 13*) is a useful aid to planning the whole process and ensuring that complete implementation of each stage is kept under control.

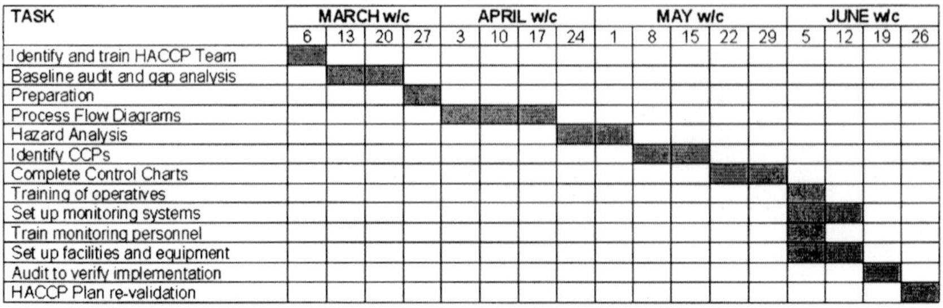

Fig. 13 Example of Gantt chart for a HACCP plan.

The team should, now, also be able to provide the rest of the organisation with a picture of what the final system will look like. This will include an indication of the documentation format, how it will link with other pre-requisite systems such as sanitation programmes or the quality management system, and an indication of the people who will be involved as the project gets underway.

When the project plan has been completed and authorised, the HACCP team can move on to key stage 2 with the knowledge that the required foundations for an effective HACCP system have been identified and will be built into the overall food safety system.

Section 3
HACCP in Practice

This section is divided into four parts:

3.1   Preparation for the HACCP plan development.
3.2   Applying the principles (the HACCP study).
3.3   Implementation of the HACCP plan.
3.4   Maintenance of the HACCP system.

Throughout this section we will be using a case study to illustrate how a fictional medium-sized manufacturing company might set about developing a HACCP system. In Appendix A a fully detailed HACCP plan is provided for the same fictional company, whereas within the text only partial details will be given as an illustration of the points being made. This fictional company is a manufacturer of a range of frozen cheesecake products.

We will be concentrating on the practical application of the seven HACCP principles as defined by Codex (1997b).

As a reminder:

Principle 1: Conduct a hazard analysis.
Principle 2: Determine the critical control points (CCPs).
Principle 3: Establish critical limit(s).
Principle 4: Establish a system to monitor control of the CCPs.
Principle 5: Establish the corrective action to be taken when monitoring indicates that a particular CCP is not under control.
Principle 6: Establish procedures for verification to confirm that the HACCP system is working effectively.
Principle 7: Establish documentation concerning all procedures and records appropriate to these principles and their application.

3.1   Preparation for the HACCP plan development

Key points

- The product description is an essential point of reference for the HACCP team as well as for future audits of the plan.
- Formulation intrinsic factors and process technologies are essential elements of product safety and must be understood.
- The process flow diagram is the basis for hazard analysis and it must contain sufficient technical detail for an effective study.
- The process flow diagram must be validated to ensure that it is accurate and representative of the process at all times.

In Section 2 we learned a little bit about the HACCP principles and also what types of preparatory and planning activities should be conducted in a business as it moves towards the use of HACCP. Here we start by seeing how a trained HACCP team will begin to develop their HACCP plan, i.e. how the HACCP study is carried out. The steps detailed in *Fig. 14* show that at this stage the first five HACCP principles are being used. Principle 6 includes validation, which is done at the end of the study.

As these steps are followed systematically, questions are asked and the answers are recorded on documents that are compiled to produce the HACCP plan. The HACCP plan is the main reference document within the HACCP system and it consists of several essential elements that are developed at this time, the process flow diagram and the HACCP control chart. It is usually kept together with other developmental documentation such as hazard analysis charts, CCP decision records, product description information and HACCP team details. There are no hard and fast rules dictating how the management and organisation aspects of the HACCP study must be done. Here we provide an example of a best practice approach.

3.1.1 Terms of reference

At the start of a HACCP study the team will confirm:

The scope of the hazard analysis

Will all three hazard groups (biological, chemical and physical) be considered at the same time? Some HACCP teams prefer to carry out a study on just one hazard group at a time. This can be simpler for inexperienced teams but it does mean that they usually have to go back and do it all over again for the other two groups once they have finished the first. It can be useful if expertise has to be bought in from outside the company, for example use of a consultant microbiologist for biological hazard identification.

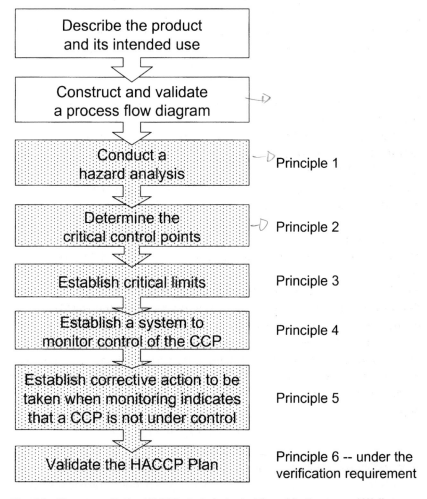

Describe the product and its intended use	
Construct and validate a process flow diagram	
Conduct a hazard analysis	Principle 1
Determine the critical control points	Principle 2
Establish critical limits	Principle 3
Establish a system to monitor control of the CCP	Principle 4
Establish corrective action to be taken when monitoring indicates that a CCP is not under control	Principle 5
Validate the HACCP Plan	Principle 6 -- under the verification requirement

Fig. 14 Key stage 2: the HACCP study (adapted from Mortimore and Wallace 1998).

The status of the prerequisite programmes

Whilst a gap analysis will usually be well underway by the time the HACCP team is ready to start the study, it is helpful to confirm understanding of which hygiene programmes are operating well and what work has to be done in this area.

The structure of the HACCP system

The team members need to have a clear understanding of the structure of the HACCP system, i.e. whether the HACCP study approach will be generic or modular – will one product or a range of products be considered. This should have already been determined in the planning stage.

The start and end points of the study

Linear HACCP studies generally start with raw materials and end with the finished product. However, modular HACCP plans need to be added together to cover the entire operation. It is important to clearly identify the start and end points of each individual HACCP study to ensure that every step is included.

Case study: Frozen cheesecake HACCP study (see Appendix A for full case study)

- All three types of hazards are being considered.
- Good manufacturing practices such as personnel hygiene and cleaning schedules are in place throughout the factory.
- The HACCP study approach is modular (six modules) and it covers a range of five different cheesecake products.

Once these terms of reference have been decided and discussed the actual HACCP study begins. Referring to *Fig. 14*, the first step in the process is concerned with making sure that the HACCP team truly understands the products being studied. This is vitally important in being able to identify likely hazards, i.e. what types of biological, chemical or physical hazards might be found in the product, where might they come from and would they survive and grow in the product.

3.1.2 Describe the product and intended use

This is normally done by looking at all the information available, including the product specification. Often findings are recorded, resulting in a brief document of about two pages. The purpose is to ensure that all HACCP team members have a good understanding of the product and process to be studied. The product description stage covers a review of the potential use of the products, the raw materials and technologies used, the main types of hazards to be considered and in general the control measures needed. This document is a vital point of reference not only for the HACCP team but also for subsequent auditors of the HACCP plan.

When designing food safety into a product it is essential to consider product formulation and the process technologies involved. These criteria are normally discussed by the HACCP team at this stage.

Product formulation

To understand how the raw materials and product formulation affect safety, the intrinsic factors of the product need to be understood. Food safety intrinsic factors are the compositional elements of the product which can affect microbial growth and, therefore, product safety. The main intrinsic factors to be considered in foodstuffs are pH and acidity, preservatives, water activity (α_w) and the ingredients themselves.

pH and acidity

- Acidity is one of the most important factors in food preservation in that it will prevent the growth of food-poisoning and food-spoilage organisms. It has been used traditionally in products such as yogurt and pickled vegetables.
- Acidity itself may be a factor in determining the organoleptic properties of a product but pH measurement is a more important factor in assessing food safety because the growth and survival of micro-organisms is usually based around the pH scale.
- The optimum pH for the growth of most micro-organisms is around neutrality (pH7) but many also grow in the range of pH4 to 8.
- A small number of bacteria may grow at pH<4 or pH>8. Those growing at pH<4 are not normally associated with food poisoning but as their growth may raise the pH to a level where pathogens could grow, it is essential that such implications are considered in the course of the HACCP study.

- Yeasts and moulds will grow at pH levels lower than 4.
- Certain micro-organisms may survive at pH values outside their growth range.

● For example:
Bacterial spores present in a low-pH raw material are unable to grow but if mixed with other raw materials to make product of higher pH they may germinate and grow to danger levels.

Preservatives

- Chemical preservatives can be added to foodstuffs to control microbial growth, subject to legal limits, and these are widely used in many products.

● For example:
Nitrites in cured meat products, sorbates in bread and cakes, meta-bisulphites in drink products.

- The traditional method of the smoking of food to preserve it also acts as a preservative by inhibiting microbial growth. This is due to the chemical compounds present in the smoke.

Water activity

Water activity (a_w) is a measure of the water available in a food for microbial growth and chemical reaction. As micro-organisms require water to grow it is possible to inhibit growth by making water unavailable to them. This can be achieved by:

- Adding solutes such as sugar and salt to a food. Sugar inhibits growth by the reduction of water activity and osmotic pressure, as in the production of jam, while salt reduces water activity and interferes with the bio-chemical processes in microbial cells, as in dry cured meat products.
- Dehydration, or the removal of water by drying foods, as in the conversion of milk to milk powder.

Ingredients

The intrinsic properties of ingredients should be assessed individually and after interaction with each other:

- The ingredients should be examined closely and any hazards associated with them considered in their own right as well as after the ingredients are mixed.

- For example:
  Preservatives and acid present in a syrup used to flavour a milk shake will stop microbial growth in the syrup itself but will have little or no effect after dilution in the mixed drink.

- Allergenic ingredients such as nuts need to be controlled through careful handling and segregated storage and processing to eliminate cross-contamination to other materials.

Case study: Frozen cheesecake HACCP study

The cheesecake in this study has a pH of 5.3–5.5 in the cheese layer and a water activity of >0.90. These will have limited or no effect on the growth of pathogenic micro-organisms, however, as the product will be frozen.

- A wide range of raw materials are used, some of which may contain pathogens and are added after thermal processing, e.g. chocolate flakes.
- Presence of allergens through use of nuts needs to be addressed.

Process technologies

The technologies used in the manufacture of a food product have different effects on its safety; it is therefore essential that each one of them is examined closely in the HACCP study. Here are some examples of process technologies and their effects on potential hazards:

Thermal processes such as heating and cooling can lead to hazards if not properly controlled.

In *freezing processes* the holding stages before freezing and length of time to freeze may be significant in allowing time for an increase in the microbial population. Cross-contamination risks are also important if the food is to be consumed without any further processing or cooking.

In *fermentation processes* the culture system needs to be controlled and identified accurately to avoid growth of undesirable micro-organisms.

Where *irradiation* is used mishandling of the food before the irradiation process may result in toxins which would not be eliminated by the irradiation process.

The *packaging system* used must also be considered as it may have an effect on product safety. This is particularly important in the case of aseptic packaging and canned products.

⬤ For example:
Thermally processed low acid canned foods will only be safe if the can seams are secure.

Case study: Frozen cheesecake HACCP study

- The product is baked and then frozen and the risk of surviving spores growing during storage is eliminated.
- There is no subsequent processing by the consumer and after thawing the shelf life is limited. There is small potential for consumer abuse.

3.1.3 Construct and validate a process flow diagram

The next step is the construction of the process flow diagram, which will be used as the basis of the hazard analysis. This is a stepwise description of the entire process. It can be done as one diagram covering the entire process or as a series of smaller diagrams, where the modular approach is being used. It must contain sufficient technical detail for the team members to be able to follow each step of the process from the delivery of the raw materials to the delivery of the end product.

The process flow diagram should include data such as:

- Details of all raw materials and packaging.
- All process activities.
- Storage conditions.
- Temperature and time profiles.

- Transfers within and between production areas.
- Equipment/design features.

It is generally useful to include as much detail as possible, although engineering drawings and symbols may cause confusion and are best avoided. It is also helpful if each step of the diagram is numbered as this can be used for cross-referencing at the hazard analysis stage.

In the modular approach it is essential that the start and finish points of each module have been well defined so that no steps are missed out accidentally and the process flow diagrams can be added together to show a picture of the whole operation. The benefits of using a modular approach include:

- The ability to include more detail within each modular diagram without the diagram appearing over-complicated.
- Operatives from each unit of operation feeling greater ownership.

Case study: Frozen cheesecake HACCP study

A modular structure consisting of six modules which fit together to cover the entire production process (see *Fig. 18* in Appendix A).

The completed process flow diagram must then be checked for accuracy by following the process in action and comparing each step with the diagram. During this validation process it is often found that steps have been missed and also on occasions that production operatives use 'short cuts' which they consider to be acceptable but which may form the basis for potential hazards. This is more likely in a highly manual operation than in an auto-mated process. Varying practices may be carried out by different operatives doing the same job; it is therefore necessary to check for accuracy on several shifts. It is rare that a completed process flow diagram will not require modification following on-site confirmation, so the importance of doing this *prior* to the hazard analysis stage cannot be stressed enough.

3.2 Applying the principles

Here we give a consideration to how each of the seven principles of HACCP may be applied and consider the key points for each.

3.2.1 Principle 1: Conduct a hazard analysis – What can go wrong?

> **Key points**
>
> - Hazard analysis of the process is where the team members systematically analyse each raw material and step of the process and identify and analyse all potential hazards and their control mechanisms (measures).
> - Hazards are biological, chemical and physical in nature.
> - Hazard analysis must be based on sound science.
> - Assessment of the likelihood of occurrence and severity of outcome is an essential part of hazard analysis and it should employ all sources of information available.
> - Hazards are considered to be significant if they are likely to cause harm to the consumer.
> - Significant food safety hazards are managed through a HACCP system. Non-significant hazards are managed through prerequisite hygiene programmes.
> - Control measures are specific for each hazard and can be either process steps or activities.

Hazard analysis is the part of the HACCP study where the team looks at each step of the process, identifies the hazards likely to be present, evaluates their significance and ensures that adequate measures for their control are in place.

HACCP was conceived as a means to control food safety hazards. Codex (1997b) refers to the identification of hazards which are 'of such a nature that their elimination or reduction to safe levels is essential to the production of safe food' and considers the definition of a hazard to be:

> 'A biological, chemical or physical agent in, or condition of, food with the potential to cause an adverse health effect.'
>
> (Codex, 1997b)

In other words, a hazard is considered to be significant if it is likely to cause harm to the consumer unless it is properly controlled. All significant hazards are managed through HACCP whereas non-significant hazards are controlled by other systems.

● For example:
  Unpasteurised milk may carry pathogenic micro-organisms that are
  significant safety hazards whereas an apple pip in a pie, undesirable
  as it may be, will not cause injury or illness.

It is essential that HACCP team members are able to understand what
constitutes a significant hazard. This involves considering each potential
hazard in turn and attempting to answer the questions:

● Could this hazard occur in the raw materials, process or finished
  product?
● Would it cause severe harm to the consumer, i.e. through serious
  injury or illness?

To answer these questions it is important that the appropriate, experienced
personnel are consulted, otherwise the resulting HACCP system could be
unsound.

Hazards may be biological, chemical or physical contaminants. They may
originate from the raw materials, the packaging, the process and handling
in the food chain or the environment.

Biological hazards

These occur in the form of pathogenic micro-organisms and they present
the biggest danger to consumers in many product groups. Pathogenic
micro-organisms exert their effect either directly through growing in or
contaminating food products and being ingested (foodborne infection), or
indirectly by forming toxins (food poisoning). In both cases the illnesses
may be serious, even fatal.

Pathogenic bacteria are extremely diverse in their nature and they grow in
many different environments.

● For example:
  Bacillus cereus forms heat resistant spores that can only be eliminated by severe heat treatment.
  Listeria monocytogenes can grow slowly at chill temperatures. However it is easily destroyed by cooking.
  ⇀ *Clostridium botulinum* requires the absence of oxygen to grow, e.g.
  canned foods, and produces a deadly toxin.

Staphylococcus aureus and *Bacillus cereus* can form toxins in food under the right conditions.

Salmonella can infect in low doses, particularly in high fat products but is easily destroyed by cooking.

Other pathogenic micro-organisms include viruses (e.g. Norwalk), toxigenic fungi (e.g. *Aspergillus*), and protozoan parasites (e.g. *Cryptosporidium parvum*).

Table 1 shows the profile of some of the more widespread pathogenic micro-organisms, how they can find their way into food and their optimum growth conditions. Use of information such as this and additional more detailed data are essential to HACCP teams as they compare the information gathered during the product description stage with the growth conditions needed by these pathogenic micro-organisms. The ability to interpret such microbiological data is important for products that are sensitive to microbial contamination and growth; consideration should therefore be given to the inclusion of an experienced microbiologist in the HACCP team.

Table 1 Pathogen profiles (adapted from Mortimore and Wallace 1998).

Organism	Sources	Associated foods	Optimum growth characteristics
Bacillus cereus	Soil, cereal crops, dust, vegetation, animal hair, fresh water and sediments	Spices Cereal ingredients	Aerobic 30–40°C pH 6.0–7.0 α_w 0.995
Campylobacter jejuni	The intestinal tract of animals	Poultry, meat, untreated water and inadequately pasteurised milk	Microaerophilic 42–43°C pH 6.5–7.5 α_w 0.997
Clostridium botulinum	Spores found in soil, shores, intestinal tract of fish and animals, deposits of lakes and coastal water	Can appear in all foods	Obligate anaerobic 25–30°C pH 7.0 α_w 0.99–0.995
Clostridium perfringens	Soil, dust, vegetation, the intestinal tract of humans and animals	Raw, dehydrated and cooked food	Aerobic 43–47°C pH 7.2 α_w 0.995

Contd

Table 1 *Contd.*

Organism	Sources	Associated foods	Optimum growth characteristics
Listeria monocytogenes	Soil, silage, sewage, faeces of healthy humans and animals	All food processing environments	Faculatative aerobic 37°C pH 7.0 α_w 0.998
E coli 0157:H7	The small intestine	Undercooked ground beef, raw milk, raw produce, infected fruit juice	Faculatative aerobic 35–40°C pH 6.0–7.0 α_w 0.995
Salmonella spp.	The intestinal tract of humans and animals, sewage	Pork, poultry, eggs, raw milk, water, shellfish	Faculatative aerobic 35–43°C pH 7.0–7.5 α_w 0.99
Shigella spp.	Hands soiled with faeces, flies	Water, milk, salads, processed potato, cooked rice, hamburgers	Aerobic 35–43°C pH 5.5–7.5
Staphylococcus aureus	Mucous membrane and skin of warm blooded animals and humans	All cooked foods	Anaerobic 37°C pH 6.0–7.0 α_w 0.98
Vibrio parahaemolyticus	Inshore warm coastal waters	Shell fish and fish	Aerobic 37°C pH 7.8–8.6 α_w 0.981
Aspergillus (aflatoxins)	The environment	Nuts, oilseeds	33°C pH 5.0–8.0 α_w 0.98–>0.99
Viruses (e.g. Norwalk, Enterovirus Hepatitis A)	Infected food handlers, sewage, contaminated water	Bivalve molluscs Passive transfer to any ready-to-eat foods	Viruses do not grow in foods

In summary, micro-organisms have basic needs related to:

- Optimum growth temperature.
- Moisture.
- Optimum acidity.
- Food source.

With time to grow in the right conditions they will be a problem. Preventative control measures must therefore be based around elimination of these basic needs.

Chemical hazards

Chemical contamination of foodstuffs can occur via the ingredients, at the time of their production or during distribution/storage and the effect on the consumer can be long term (e.g. carcinogenic), short term (e.g. allergic reactions) or teratogenic (e.g. BSE in animals).

- Some examples of chemical contamination:
  In raw materials: pesticides/herbicides, toxins (natural or microbial), allergens, antibiotics, hormone residues, heavy metals.
  During process: cleaning agents, lubricants, refrigerants, pest control chemicals, toxins, allergens.
  From packaging: plasticisers and additives, ink, adhesive, metal leaching from cans.

The HACCP team needs to review any toxic chemicals on the premises as well as considering likely contaminants in raw materials and packaging.

Physical hazards

These are foreign bodies or matter that can contaminate a foodstuff at any time during production. Strictly speaking they are only significant safety hazards if they are likely to cause injury or a health risk to the consumer; otherwise they should be considered in terms of quality, wholesomeness or legality and managed through hygiene and quality prerequisite control programmes. A HACCP study can identify all potential foreign matter and it can be extended to include quality and legality issues, but the controls should be clearly separated from those that are critical to food safety.

Foreign material items are considered as food safety hazards themselves if they fall under the following categories:

- Items that are sharp and can cause pain and injury, e.g. wood splinters, glass fragments.
- Items that can cause severe dental damage, e.g. metal, stones.
- Items capable of causing choking, e.g. bones or plastic.

Another reason for managing foreign matter contamination is that it can act as a vehicle for micro-biological cross-contamination. An example of this is a fly in a fresh cream cake where the transfer of pathogenic micro-organisms from the fly to the cake would present the hazard, not the fly itself.

Useful sources of information when carrying out a hazard analysis in all three groups are:

- Published information in books, scientific journals and on the Internet.
- Consultants/specialists.
- Research organisations.
- Suppliers/customers.

All information must be researched and evaluated thoroughly before any conclusions are drawn.

The hazard analysis is carried out by following the process flow diagram and discussing the hazards that may potentially occur at each raw material and process step. This is often done by a brainstorming approach and capturing output on a flip chart. The source or cause of each hazard should also be documented as this provides the team with a better understanding of how to control the hazard.

Risk assessment is an important part of hazard characterisation that helps to determine the significant hazard. In terms of HACCP, a risk is defined by Codex (1998a) as:

'A function of the probability of an adverse health effect and the severity of that effect consequential to a hazard(s) in food.'

In other words, the probability or likelihood that a severe health effect will be realised. It is important at this stage that the HACCP team has complete knowledge of their raw materials and processes when deciding whether a hazard will realistically occur, but also that they are aware of their own limitations when making judgements.

The topic of risk, its analysis and assessment is still being discussed widely within governments and academia (see Epilogue). For the purpose of HACCP at business level HACCP teams will be making a qualitative evaluation during the hazard analysis process, i.e. is the hazard significant enough to require management through the HACCP plan or can it be managed through pre-requisite programmes and/or through the quality system?

Formal risk assessment is a quantitative, global process where a numerical degree of risk can be calculated for a particular hazard (Sperber 2001 in press). It is undertaken at government level, usually with involvement of academia and industry.

Qualitative hazard analysis is the responsibility of the individual HACCP team and should include the following where possible:

- The likely occurrence of hazards and the severity of their adverse health effects.
- Qualitative and/or quantitative (if data is available) evaluation of the presence of hazards.
- Survival or growth of pathogenic microorganisms.
- Presence of toxins, chemical or physical agents.
- Conditions that may lead to the above.

When all the significant hazards have been determined the HACCP team must consider what required control measures must be in place for each identified hazard. These may already be in operation at the particular step examined or may occur later in the process, but they need to be re-evaluated to ensure that they are adequate. Additional control measures may also be required and these will have to be planned and developed.

Experienced HACCP teams may decide to review the control measure options as each hazard is identified. This can lead to long discussions which divert the focus away from the objective of hazard identification and introduces the possibility of missing out potential hazards. It can be simpler to identify all the significant hazards before starting to consider control measures.

The sources of biological hazards are extremely diverse and they have to be controlled through a variety of control measures. These may be at any specific point of the food supply chain and it is important that each control measure is applied at the correct point to ensure that it is effective. *Table 2* shows some examples of biological hazards and how they can be controlled.

Many chemical hazards are controlled through prerequisite programmes such as supplier ingredient control and good manufacturing practices but it is often found that where nut-containing products are manufactured on a non-dedicated line the risk of allergen cross-contamination needs to be

Table 2 Examples of control measures for biological hazards (adapted from Mortimore and Wallace 1998).

Biological hazard	Control measures
Vegetative pathogens e.g. *Salmonella, Listeria monocytogenes, E coli*	**Raw materials** Lethal heat treatment during process Specification and surveillance (SQA) Effective supplier process and testing Certificate of analysis Temperature control **Cross-contamination** Intact packaging Pest control Secure building (no roof leaks, ground water) Logical process flow (segregation of people, clothes, equipment etc., direction of drains) Intrinsic factors, pH, α_w etc.
Spore formers e.g. *Clostridium botulinum, Bacillus cereus*	**Raw materials** Specification and surveillance (SQA) Effective supplier process and testing Certificate of analysis Lethal heat treatment during process Temperature control – prevention of spore outgrowth, e.g. after heat treatment *Cross-contamination* Intact packaging Pest control Secure building (no roof leaks, ground water) Logical process flow (segregation of people, clothes, equipment etc., direction of drains) Intrinsic factors, pH, α_w etc.
Heat-stable pre-formed toxins e.g. *Staphylococcus aureus, Bacillus cereus*, emetic toxin	**Raw materials** Specification for organism and/or toxin and surveillance (SQA) Effective supplier process and testing Certificate of analysis **People** Handwash procedures Covering cuts/wounds etc. Occupational health procedures Management control of food handlers **Build-up during process** Control of time that ingredients, intermediate and finished products are held within the organism's growth temperature range Design of process equipment to minimise dead spaces 'Clean as you go' procedures Control of rework loops Validation studies on maximum length of production run without cleaning
Mycotoxins e.g. patulin, aflatoxin, vomitoxin	SQA control of harvesting and storage to prevent mould growth and mycotoxin formation Heat treatment during process to destroy mould Controlled dry storage

Table 3 Examples of control measures for chemical hazards (adapted from Mortimore and Wallace 1998).

Hazard	Control measures
Cleaning chemicals	Use of non-toxic, food compatible cleaning compounds Safe operating practices and written cleaning instructions Separate storage for cleaning reagents Designated, covered containers for all chemicals
Pesticides, veterinary residues and plasticisers in packaging	Specification to include suppliers' compliance with maximum legal levels Verification of suppliers' records Annual surveillance programme of selected raw materials
Toxic metals/PCBs	Specifications and surveillance where appropriate
Chemical additives e.g. nitrates, nitrites	*As contaminants* Specifications and surveillance where appropriate (SQA) *As additives* Safe operating practices and written additive instructions Special storage in covered, designated labelled containers Validation of levels through usage rates, sampling and testing
Allergens/food intolerance	Awareness of the potential allergenic properties of certain ingredients and training of the workforce. Special consideration given to adequate labelling, production scheduling and cleaning, segregation or cross-contamination controls, dedicated equipment, and to the control of rework Final rinse water testing should be considered as a verification method where possible

managed through a HACCP system. *Table 3* shows some examples of chemical hazards and how they can be controlled.

Physical hazards are mainly foreign bodies that can be controlled by pre-requisite programmes, as shown on *Table 4*.

More than one control measure may be needed to control an identified hazard. Also, more than one hazard at a process step may be controlled by a single control measure. Control measures should not be confused with monitoring – the control measure actually controls the hazard by preventing, eliminating or reducing it to an acceptable level, while the monitoring activities simply provide an indication that control is being achieved.

Table 4 Examples of control measures for physical hazards (Mortimore and Wallace 1998).

Hazard	Control measures
Intrinsic physical contamination of raw materials e.g. bone in meat/ fish, fruit stones, stalks, pips, nutshells	*Liquids* Filtering, magnets, centrifugal separation *Powders* Sifting, magnets, metal detection
Extrinsic physical contamination of raw materials e.g. glass, wood, metal, plastic, pests	*Flowing particles, e.g. nuts, dried fruit, IQF fruit and vegetables* 100% inspection – electronic or human Screening, sifting, magnets, metal detection Washing, stone and sand traps Air separation, flotation, electronic colour sorting *Large solid items, e.g. carcasses, fish, cabbages, cauliflowers, frozen pastry, packaging* X-ray detection, metal detection De-boners, visual inspection, electronic scanning
Physical process cross-contaminants e.g. glass, wood, metal, plastic, pests	Elimination of all glass except lighting which should be covered – light breakage procedure Glass-packed products – glass breakage procedures, inversion/washing/blowing of glass packaging before use Exclusion of all wooden materials such as pallets, brushes, pencils, tools from exposed product areas Segregation of all packaging materials Equipment design – preventative maintenance Avoidance of all loose metal items – jewellery, drawing pins, nuts and bolts, small tools Metal detection – sensitivity appropriate for the product, calibrated (3-monthly) and checked (hourly), ferrous, non-ferrous and stainless steel; fail-safe divert systems; locked reject cases, traceability Avoidance of all loose plastic items – pen tops, buttons on overalls, jewellery Breakage procedure in place where brittle plastic is used *Pest control programme* Prevention (facility design, avoidance of harbourage areas, waste management, ultrasonic repellents) Screening/proofing (strip curtains, drain covers, mesh on windows, air curtains, netting) Extermination (electric fly killers, poisoning, bait boxes, traps, perimeter spraying, fogging)
Building fabric	*Design and maintenance* Regular inspection

Case study: Frozen cheesecake HACCP study

Several biological, physical and chemical hazards are identified in the raw materials and process steps, requiring a wide variety of control measures.

- For example:

  A raw material (chopped hazelnuts) is considered to have associated aflatoxin and allergen contamination hazards which are controlled through effective supplier management. In addition, allergen cross-contamination issues will need to be considered during the process. The allergen issue is two-fold:

  1. Consumers may be allergic to hazelnuts and wish to avoid eating any products containing them. Effective labelling of hazelnut-containing products and prevention of cross-contamination to other products will be important here.

  2. People who can safely consume hazelnuts may be allergic to other nuts, e.g. peanuts, so cross-contamination at the nut supplier must be considered.

  At the baking step, survival of vegetative pathogens is considered and this is controlled through ensuring that the correct heat process occurs.

All the identified hazards and their control measures for each raw material and process step are usually recorded on a hazard analysis worksheet or chart. An example is given in *Table 5*. This particular worksheet includes space for CCP identification which is discussed below.

Table 5 Example of a worksheet for hazard analysis and CCP identification.

Process step	Hazard	Control measures	Q1	Q1a	Q2	Q3	Q4	CCP	Justification

3.2.2 Principle 2: Determine the critical control points (CCPs) – At what stage in the process is control essential?

> ### Key points
>
> - The identification of CCPs relies on expert judgement.
> - To identify CCPs all raw materials, process steps and their control measures can be assessed using the CCP decision tree as a tool.

A critical control point is a step at which 'control can be applied and is essential to prevent or eliminate a food safety hazard or reduce it to an acceptable level' (Codex 1997b). CCPs relate to control of significant food safety hazards only.

As discussed earlier, control points (CPs) relating to quality or legal issues are managed by other programmes. It is essential that this relationship (CCPs versus CPs) is clearly understood by the HACCP team members in order to ensure that only safety points are determined as CCPs. There is sometimes a tendency to designate too many CCPs 'to be on the safe side'. This can undermine the system, losing credibility, and can make imple-mentation more difficult to manage because resource is spread too thinly. On the other hand too few CCPs may result in the production of unsafe food.

If there is any difficulty in telling the difference between CPs and CCPs, ask the simple question: If control is lost, is it likely that a health hazard will occur? If the answer is 'yes' then the point must be managed as a CCP. If the answer is 'no', i.e. food safety is not necessarily compromised, then the point may be managed as a CP.

Identification of CCPs can be carried out by using tools such as decision trees, an example of which is given in *Fig. 15*. The most widely used decision trees are those published by Codex (1997b) and NACMCF (1997), though variations exist (ILSI 1997; Mortimore & Wallace 1998). Raw materials can go through the Codex style decision tree but the wording does not always lend itself to a raw material situation. Where companies are introducing new raw materials on a frequent basis and/or undertake a lot of new product development it can be helpful to evaluate those hazards separately using an alternative raw material decision tree (ILSI 1997, Mortimore and Wallace 1998).

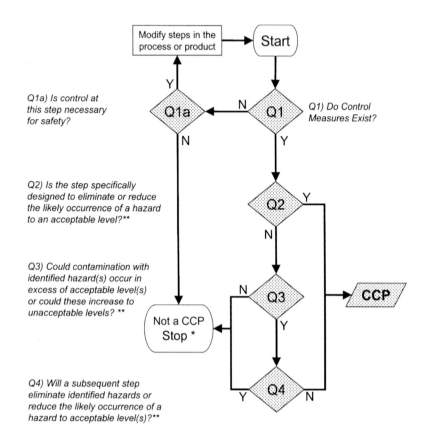

Q1a) Is control at this step necessary for safety?

Q1) Do Control Measures Exist?

Q2) Is the step specifically designed to eliminate or reduce the likely occurrence of a hazard to an acceptable level?**

Q3) Could contamination with identified hazard(s) occur in excess of acceptable level(s) or could these increase to unacceptable levels? **

Q4) Will a subsequent step eliminate identified hazards or reduce the likely occurrence of a hazard to acceptable level(s)?**

* Proceed to the next identified hazard in the described process

** Acceptable and unacceptable levels need to be determined within the objectives in identifying the CCPs of the HACCP plans

Fig. 15 Example of decision tree for process steps (adapted from Codex 1997b).

The essential skill needed for CCP identification is thorough knowledge of the product, the process, the identified hazards and the measures for their control. The information collated during the product description and hazard analysis stages is used by the team to determine the CCPs, while decision trees are helpful in providing a structured approach. Expert judgement must always be used and specialist advisers can be usefully drafted into the team at this stage if necessary.

Decision trees consist of a number of questions that are applied to each identified hazard listed in the hazard analysis chart, as follows.

Q1. Do control measures exist?

This is answered by referring to the control measure data which was documented on the hazard analysis chart and also by making sure that those control measures really are in place and operational within the business.

If the answer to Q1 is 'Yes' move to Q2. If the answer is 'No' move to Q1a.

Q1a. Is control at this step necessary for safety?

If the answer is 'No' (for example, a control measure may be in place further along in the process that will control the hazard) move to the next process step or hazard.

If 'Yes' a modification in the process (e.g. add a sieve or increase a process temperature) or product (e.g. reformulate to reduce the pH or add preservatives) must be implemented to ensure that control measures can be built in. When a suitable control measure has been identified go back to Q1 (the answer will now be 'yes') and progress through the tree.

Q2. Is the step specifically designed to eliminate or reduce the likely occurrence of a hazard to an acceptable level?

This question aims at finding out whether the step under consideration is effective at controlling food safety. The thing to remember when asking this question is that it refers to the process step and not the control measure. If the control measure is considered at this stage the answer will always be 'Yes' and consequently the step may be wrongly labelled a CCP.

- For example:
  Milk must be heat-treated at a specific temperature for a specific time in order to eliminate all vegetative pathogens that may be present. This process step of pasteurisation is designed to reduce the likely occurrence of a hazard to an acceptable level and is therefore a CCP.

If the answer to Q2 is 'Yes' the process step is a CCP. Start the decision tree again for the next process step or hazard. If the answer is 'No' move to Q3.

Q3. Could contamination with identified hazard(s) occur in excess of acceptable level(s) or could these increase to unacceptable levels?

To answer this question the team should use the information recorded on the hazard analysis chart and their expert knowledge of the process and its environment. The issues to be considered here include:

- Time and temperature conditions.
- The production environment (design, hygiene, maintenance).
- Cross-contamination from personnel, another product or raw material.
- Acceptable levels for significant hazards.

Expert advice should be sought when necessary.

If the answer to Q3 is 'Yes' proceed to Q4. If the answer is 'No' move to the next process step or hazard.

Q4. Will a subsequent step eliminate identified hazards or reduce the likely occurrence of a hazard to acceptable level(s)?

This question is designed to acknowledge the presence of any hazards that will be removed by subsequent steps later in the process or by the consumer.

If the answer to Q4 is 'Yes' move to the next process step or hazard. If the answer is 'No' the process step is confirmed as a CCP. Start the decision tree again for the next process step or hazard.

In addition to expert judgement, the identification of the CCPs in a process also requires some common sense. Panisello and Quantick (2000) suggest that it is unlikely that a CCP would be appropriate in the following circumstances.

Where a hazard cannot be controlled

Quite often cross-contamination issues would come into this category. By following all prerequisite programmes, environmental cross-contamination should be reduced but it would be wrong to say that it could be guaranteed as eliminated. Where human operatives are involved there is a heavy reliance on behaviour, which is unpredictable.

Where there is no possibility of establishing a scientifically based critical limit

Again, where prerequisite hygiene issues such as 'cleanliness of equipment' are cited as being the control measure. It is difficult to set a true value on the cleanliness required for food safety control.

Another example might be cooling in a spiral chiller. At what point during the hour or so of cooling does the temperature drop become critical?

Where the step cannot be monitored

For example in the spiral chiller situation above, or in a small retail or catering situation where they are unlikely to have metal detectors to control metal contamination.

Case study: Frozen cheesecake HACCP study

The steps identified as CCPs are outlined in Table 9 in Appendix A. These include incoming raw materials (chocolate flakes and chopped hazelnuts) where the control measures involve supplier control, and also the process steps of *baking* to control vegetative pathogens, *scanning* packed product to control the risk of allergenic product in unlabelled packaging and *metal detection* for effective removal of any product containing metal.

The answers to all the questions in the decision tree together with any justifying comments made by the team are usefully recorded on a hazard analysis worksheet or chart. This is a good reminder of why particular process steps are or are not identified as CCPs, particularly when the system is being audited or updated at a later time.

Once identified, the CCPs should be documented on a HACCP control chart (also known as a HACCP worksheet). An example is given in *Table 6.*

Table 6 Example of a HACCP control chart (or worksheet).

Raw material/ process step	CCP no	Hazard to be controlled	Control measure	Critical limits	Monitoring			Corrective action	
					Procedure	Frequency	Responsibility	Procedure	Responsibility

3.2.3 Principle 3: Establish critical limit(s) – What criteria must be met to ensure product safety?

Key points

- Critical limits are the criteria that differentiate between 'safe' and potentially 'unsafe'.
- Critical limits are defined by regulations, safety standards and scientifically proven values.
- They are measurable parameters that can be determined and monitored through testing and observation.
- Operational limits are often set at more stringent levels to provide a buffer or action zone for process management.

Once all the CCPs have been identified the team has to decide the criteria that distinguish between 'safe' and potentially 'unsafe' for each CCP. These are represented by defined parameters called critical limits. When the product falls outside these limits the CCP is out of control and a safety hazard may be present. Critical limits may be defined by regulations, company safety standards or scientifically proven values.

The Codex (1997b) definition of a critical limit is 'a criterion which separates acceptability from unacceptability'; in other words 'critical limits are the safety limits that must be met for each control measure at a CCP'.

The HACCP team must fully understand the criteria that govern safety at each CCP and the factors that are associated with them in order to decide the appropriate critical limits. These factors are related to the type of hazard that the CCP is designed to control and must be measurable parameters that can be determined and monitored through testing or observation. Criteria often used include measurements of temperature, time, moisture level, pH, a_w, and available chlorine and sensory parameters such as visual appearance and texture (Codex 1997b). Critical limits must be validated, i.e. the criteria specified must control the identified hazard.

In addition to the critical limits it is usual to have operational limits which provide a buffer or action zone for process management. These are designed to allow for a certain amount of deviation in normal process operation while ensuring that food safety is not compromised.

● For example:

If in a heat process the critical limit is 72°C for 2 minutes, the operating limit of 75°C for 10 minutes may be set.

Once defined, the critical limits are recorded on the HACCP control chart.

Case study: Frozen cheesecake HACCP study

● The critical limits are validated through testing, e.g. at the baking step a validation study is done to confirm that the heat process of 140°C for 55 minutes will achieve the required minimum core temperature of 72°C in all areas of the oven.

● In the two raw materials that have been identified as CCPs the critical limit is only purchasing from suppliers approved through an effective supplier quality assurance/vendor assurance (SQA/VA) programme. Some people consider that SQA/VA programmes are prerequisites and should not be identified as CCPs. However it is useful to know which raw materials are critical in terms of the hazards they might introduce and few companies have SQA/VA programmes that are so well developed that this focus on critical raw materials is unhelpful.

3.2.4 Principle 4: Establish a system to monitor control of the CCP – What checks will indicate that something is going wrong?

Key points

● Monitoring is the measurement or observation required to ensure that the process is under control and operating within the defined critical limits.

● Monitoring of the CCPs is carried out through tests or observations.

● The frequency and responsibility for monitoring will be appropriate to the control measure.

● All personnel responsible for monitoring must be trained and have a clear understanding of their role.

Monitoring is the measurement or observation that the process is operating within the critical limits (or more likely the operating limits) at the CCP. The Codex (1997b) definition of monitoring is 'the act of conducting a planned sequence of observations or measurements of control parameters to assess whether a CCP is under control'. So, the reason for monitoring the CCPs is to confirm that they are working and that safe food is being produced.

The procedures for monitoring the CCPs will depend on the nature of the control measure and also on the capabilities of the monitoring device or method used. It is essential that the monitoring is able to detect loss of control, otherwise the system becomes invalid.

As already stated, monitoring should not be confused with control measures. Monitoring is the act of carrying out tests and observations to ensure that the process is under control, i.e. that control measures are working. It is a surveillance activity and provides the records that will be used later as part of the verification process.

Establishing monitoring procedures involves a number of elements, as follows.

Equipment and methods

Referring back to the previous discussion on critical limits, we stated that criteria used may include, and therefore require, measurements of:

1. Physical parameters, such as temperature, time and moisture levels. Other types of physical measurements are checking the operation of metal detection, magnets, X-ray detection, and inspecting sifters and sieves.
2. Chemical tests, such as chlorine analysis, pH, a_w. Other types of chemical tests might include pesticide residue analysis, allergen residue testing or heavy metals analysis.
3. Sensory tests, such as visual appearance and texture. Although such tests are often associated with quality criteria, visual appearance monitoring may be involved in foreign material CCPs and texture may be critical to effective heat penetration in, e.g. a canned product.

These lists are not exhaustive. Microbiological testing is not usually used for monitoring as the results are not immediately available. Real-time results are preferable as this allows fast corrective action to be taken.

The equipment used for monitoring must be:

- Accurate – it needs to be calibrated to ensure reliable results.
- Easy to use – it is not always practical to have to use a piece of complicated equipment, particularly if the monitoring is carried out by operatives in a production environment.
- Accessible – having the equipment close to the point of testing means that the test is likely to be quick in terms of providing results to the people who are involved in the process.

Monitoring procedures may involve:

- On-line systems where the critical factors are measured during the process continuously or at intervals.

  - For example:
    A temperature chart recorder (thermograph) operating continuously.

- Off-line systems where samples are taken and measurements made later.

  - For example:
    Samples of product taken for pH checks.

- Observational systems

  - For example:
    Confirming that the metal detector will detect metal when test pieces are passed through at defined intervals and that the automatic reject mechanism is synchronised to reject the correct product.

Most monitoring systems are based on traditional forms of inspection and testing which have limitations, i.e. off-line and at intervals. They are however useful to demonstrate that control has been achieved, provided that they have been designed properly. On-line sensors are being increasingly used to provide a higher level of confidence.

Frequency determination

The frequency of monitoring will depend on the nature of the CCP and it must be determined as part of the control system.

For example:
A metal detector will be monitoring the presence of metal continuously and will be verified as working at defined intervals (e.g.
hourly).
Acidity monitoring may be done by measuring the pH of each batch of
product produced.

People

The allocation of responsibility, as with setting the frequency of monitoring,
also depends on the nature of the CCP. People assigned monitoring duties
should be (NACMCF 1997):

- Familiar with the process.
- Trained in the monitoring techniques.
- Trained in HACCP awareness, i.e. so that they appreciate their role
  and its importance in relation to the business's HACCP plans.
- Unbiased in monitoring and reporting.
- Trained in the corrective action procedures, i.e. what to do when
  monitoring indicates loss of control.

Record keeping

This is worth mentioning here though it will be discussed again later under
HACCP Principle 7.

The results of monitoring activities should be recorded by the CCP monitors. Log sheets used to document all checks must contain the correct
information to ensure that the CCP is under control and they must be
regularly reviewed and signed off by a trained supervisory level authority. In
some countries where HACCP is mandatory, the monitoring records are
legal documents. CCP monitoring records can be kept as part of general
production log sheets.

In any situation it is the monitoring records which provide the evidence
that the process was under control and that food has been made in
accordance with the critical controls identified, i.e. those which ensure
safe food.

Details of the monitoring procedure, frequency and responsibility are
recorded in the HACCP control chart.

Case study: Frozen cheesecake HACCP study

- Monitoring of the raw material CCPs is carried out through SQA audits and checks against the approved supplier list and certificates of analysis.
- The baking CCP is monitored by visually checking the oven chart recorder and signing off.
- The scanning and metal detection CCPs are monitored by checking that the equipment is functioning at regular intervals.

3.2.5 Principle 5: Establish the corrective action to be taken when monitoring indicates that a particular CCP is not under control – If something does go wrong what action needs to be taken?

Key points

- Corrective actions must be properly defined to ensure that the consumer is protected and that control is regained.
- Since HACCP is a preventative system, corrective actions should be such that further deviations are prevented.
- Following a deviation there are two priorities: to deal with product produced during the deviation period and to bring the process under control.
- Responsibility for corrective action must be assigned by management.

When monitoring results show a deviation from the critical limits at a CCP, corrective action must be taken. Since HACCP is designed to prevent such deviations happening in the first place, corrective actions should be considered at two levels:

1. What needs to be done following a deviation at a CCP, i.e. corrective action.
2. Modification of the process so that further deviations are prevented, i.e. preventative action.

What needs to be done following a deviation at a CCP?

When a deviation at a CCP occurs, quick action is essential. The aim is to:

- Deal with the material produced during the deviation period, and
- Bring the process under control.

All material produced during the deviation period should be put securely on hold while the likelihood of the hazard being present in the non-complying product is considered. Where appropriate the product may be tested and the test results assessed statistically in terms of product safety. After all considerations have been made the product may be destroyed, re-worked, re-tested and released or put to alternative use, e.g. sold as animal feed.

Off-line responsibilities for this type of corrective action are usually handled by more senior personnel and the HACCP team leader may also need to be involved.

Bringing the process under control may involve stopping the line and a temporary solution being put into effect so that the line can be restarted while a permanent corrective action is sought.

For example:
The provision of a temporary off-line metal detector while the on-line detector is repaired.

Responsibility for this type of corrective action needs to be agreed with the production management who are implementing the HACCP plan. Responsibilities will often lie with the operator monitoring the CCP, who will take immediate corrective action and/or will notify the supervisor for further action.

Modification of the process so that further deviations are prevented

This involves adjusting the process or product so that control is maintained. Typical examples of such changes are increasing specified cooking time or adjusting the pH through product reformulation. This would involve adjusting the operating limits to give a bigger buffer/action zone.

Responsibilities for this type of activity will always include the HACCP team as the HACCP plan will need to be reassessed for any proposed changes and updated where necessary.

All corrective actions do need to be thought through before being listed in the HACCP plan. It is not helpful if the HACCP team simply decides that the corrective action would be to 'report to supervisor' if the supervisor is unclear as to what to do next.

Details of the corrective action and responsibility are recorded in the HACCP control chart. At this stage the HACCP control chart should be complete.

Case study: Frozen cheesecake HACCP study

- Corrective actions for the raw material CCPs include rejecting the batch.
- For scanning and metal detection the corrective action involves identifying and rechecking product since the previous satisfactory check.
- For cooking, the line manager is informed and the main corrective action is to continue cooking or re-cook, depending on when the problem is identified for the batch.
- Responsibility at each CCP lies with personnel of various levels but in many cases a management decision will need to be taken.

3.2.6 Principle 6: Establish procedures for verification to confirm that the HACCP system is working effectively – How can you make sure that the system is working in practice?

Key points

- Verification of the system is essential to ensure that all hazards can be controlled and that all controls are operating correctly.
- A HACCP plan is only valid after all the details have been checked to ensure that all control measures will control the identified hazards and that the HACCP plan is complete.
- The use of expert resource may be beneficial in ensuring the validity of the plan.
- Verification is carried out through auditing, product testing and reviewing records, procedures and practices where necessary.
- Production and monitoring equipment should also be checked for its ability to achieve what is required, i.e. that it is calibrated.

The Codex (1997b) definitions of validation and verification are:

- Validation is 'obtaining evidence that the elements of the HACCP plan are effective'.
- Verification is 'the application of methods, procedures, tests and other evaluations, in addition to monitoring, to determine compliance with the HACCP plan'.

To appreciate the importance of this principle it is worth considering the consequences in the event of getting it wrong. These will be grave, ranging from personal injury to bad publicity and possible prosecution, and will most certainly prove expensive through compensation payments, brand damage and loss of business.

The application of Principle 6 is achieved through a number of activities which broadly fall into the two categories of validation and verification. Whilst validation is a one-off activity during the plan development it does need to be repeated should there be any change in the product or process. Verification however is an ongoing activity once the HACCP plan has been implemented.

Validation

Once the study is complete the team will need to carry out validation activities to confirm that all elements of the HACCP plan will be effective before moving into implementation. Slightly confusingly, validation appears as an activity within the verification requirement in the Codex Alimentarius guidelines; however, validation is really asking: 'Does this HACCP plan ensure that the relevant hazards have been identified and can be controlled?'. It is an important task and should be done thoroughly for each CCP identified. Validation involves working back through all the HACCP principles with the aim of making sure that control criteria have been set correctly to ensure that all the significant hazards can be controlled. It is the confirmation that the control measure and critical limit will control the identified hazard, i.e. that the information in the HACCP plan will effectively manage food safety.

It may be appropriate to use expert resource from outside the organisation to cross-check the study and ensure that all relevant issues have been covered, particularly if the HACCP plan is compiled by a team with limited experience or if the study covers a type of product or technology that is new to the company.

Case study: Frozen cheesecake HACCP study

The critical limit for the elimination of *Salmonella* in the product is 72°C. To ensure that the *centre* of each unit reaches this temperature the product is cooked at 140°C for 55 minutes.

Validation: Vegetative pathogens will be destroyed by cooking the product at the baking step to a 72°C critical limit. Validation studies would confirm that cooking to 140°C for 55 minutes ensures that the product centre temperature reaches 72°C throughout the oven.

Statistical techniques can be used during validation to establish the capability of maintaining the process within specified limits.

When the HACCP team is satisfied that all the controls will control the identified hazard the HACCP plan can be implemented.

Verification

Verification is the confirmation that the control measures have been met during the process, i.e. usually once the HACCP plan has been implemented.

- For example:
  Records confirm that the cheesecake was baked for 55 minutes at 140°C.

Verification activities include auditing the HACCP system, and review and analysis of data such as CCP records to ensure compliance, microbiological and chemical product sampling and testing, review of customer complaint records and calibration of equipment. Verification is an ongoing activity.

Regular auditing

Auditing is a key verification activity and should include inspection of production records, deviations, actions taken and reviewing the practices and procedures used to control CCPs. If internal auditing is being done, i.e. by the business itself, it is important that auditing is carried out by personnel who have not been involved in this particular HACCP study or in the day to day management of the HACCP plans, as they are detached from the process and likely to be more objective in their approach. External auditing is also a verification activity and likely to be carried out by customers, government inspectors or third parties employed either by customers or the business itself.

Regular auditing provides evidence that the HACCP plan continues to be effective. The benefits of a HACCP system audit are:

- Continuing confidence in the effectiveness of the system and awareness in food safety management.
- Improving the system through the identification of weak areas.
- Providing documented evidence that food safety is managed.

Data analysis

As already indicated, the records generated by the HACCP system must be reviewed on a regular basis as part of the verification process. This ensures that the HACCP plan continues to be effective, trends can be analysed and corrective actions put into place. Given the requirement for prerequisite good hygiene practice, these records should also be reviewed where they exist.

The types of data that need regular reviews are varied and they might include:

HACCP
CCP log sheets
Test results
Process control charts
HACCP audit reports
Customer complaints

Prerequisites
Pest control records
Glass register
Housekeeping or hygiene audit reports

The frequency of reviewing these records will depend on their nature and importance and it may be daily, weekly, monthly, quarterly or annually. Analysis is best handled electronically, wherever possible, and through the use of graphs and charts it should indicate trends and provide a visual record. Statistical techniques can also be utilised to great advantage.

Table 7 summarises the likely activities when validating and verifying that each of the seven principles has been applied correctly. This is a useful checklist.

Table 7 Examples of validation and conformity verification (adapted from ILSI 1999).

HACCP principle	Validation *Evidence to demonstrate that:*	Verification *Evidence to demonstrate that:*
1. Hazard analysis.	The correct skills were in the HACCP team. The flow diagram is suitable for the purposes of the HACCP study and all the significant hazards were identified.	Validation was carried out correctly. Product safety implications of process changes are being actively considered through hazard analysis.
2. Determination of the CCPs required to control identified hazards.	All significant hazards were considered during CCP identification. There are CCPs to control all significant hazards. The CCPs are at the appropriate stages in the process.	Validation was carried out correctly. CCPs are in place in the operation. Control measures are working in practice at each CCP.
3. Specification of critical limits to assure that an operation is under control at a particular CCP.	The critical limits control the identified hazards.	Validation was carried out correctly. Operating limits are/continue to be set at appropriate levels.
4. Establishment and implementation of systems to monitor control of CCPs.	The monitoring system will ensure that the control measures at the CCP will be effective. Procedures for the necessary calibration of testing equipment are included.	Records of monitoring exist and confirm control. Statistical process control is used where appropriate. Review of monitoring records by designated person. Records of calibration exist and confirm compliance.
5. Establishment of the corrective action to be taken when monitoring indicates that a particular CCP is not under control.	Corrective actions will prevent non-conforming product from reaching the consumer. Authority for corrective actions has been assigned.	In cases of non-conformity control is regained and appropriate steps are taken to prevent unsafe product reaching the consumer. Corrective actions are recorded and actions taken by designated persons.
6. Establishment of procedures for verification to confirm that the HACCP system is working effectively.	Procedures for information gathering and compliance verification of the HACCP system have been established.	All verification procedures are defined and carried out.

Contd.

Table 7 *Contd.*

HACCP principle	Validation *Evidence to demonstrate that:*	Verification *Evidence to demonstrate that:*
7. Establishment of documentation concerning all procedures and records appropriate to these principles and their application.	Documentation covering the entire HACCP system has been established.	Documentation and record keeping covering the entire HACCP system is complete, in the correct format, properly filled out and up to date.
HACCP training	The training materials and delivery meet the objectives, i.e. that the HACCP team understood how to apply the principles of HACCP.	The appropriate people were trained correctly.

3.2.7 Principle 7: Establish appropriate documentation concerning all procedures and records appropriate to the HACCP principles and their application – How can you demonstrate (if challenged) that the system works?

Key points

- Appropriate documentation and records are needed to demonstrate the effectiveness of the HACCP system.
- Records must be kept for a length of time defined by legislation and the shelf-life of the product.

The HACCP system must be documented and records maintained to demonstrate that it is both properly established and working correctly, i.e. Principle 7 really applies across all of the other six principles. This will support due diligence (as required by UK legislation) or any other litigation proceedings. In having legal status in many countries it is important that the documents and records are of good standard, i.e. legible, with no crossing out or correcting fluid. All documents should be signed and dated. Records are essential in analysing trends, which will be needed when reviewing and improving the system.

Document control will be easier if well organised:

- Each HACCP plan can be allocated a unique reference number which is cross-referenced on all documentation relating to it. This makes it easier to track records during implementation.
- Records must be archived and kept for an adequate length of time, which may reflect legislative requirements of the country where the product is manufactured or sold and the shelf-life of the product. As a general rule, records should be kept for at least a year after the end of product shelf-life, though a certified quality management system may require this period to be extended to three years.
- Documents should be readily accessible. In some countries they are truly legal documents which the regulator can demand to see.
- Updates or revisions to any documents should be done in a controlled way, i.e. dated and authorised. Many companies keep a 'history of amendments' in order to track the development of the system.

The types of records which will be retained include:

- The HACCP plan, which will include as a minimum the process flow diagram and HACCP control chart, together with support information (e.g. the hazard analysis, HACCP team details, product description).
- History of amendments to the HACCP plan, which will demonstrate any changes carried out.
- CCP monitoring records.
- Hold/trace/recall records generated in handling deviations.
- Training records proving that personnel involved in implementing the HACCP system have been trained to do so.
- Audit records.
- Calibration records.

In smaller businesses there is often a concern about the large amount of paperwork inevitably needed for HACCP and, in this respect, Principle 7 is not a legal requirement in many European countries. It is possible that concerns have arisen because HACCP is not properly understood, and, more specifically, because the relationship between HACCP and hygiene prerequisite programmes is not understood. Records and documents are needed to confirm that the CCPs are properly identified and are under control. If the CCPs have not been identified correctly (and usually this means that they have been identified but are hidden among many other CPs which have also been classed as CCPs), then there will be a lot of monitoring and hence record keeping. If the few true CCPs have been identified then this should not really be a burden and, in fact, should help the business to track process performance.

3.3 Implementation of the HACCP plan

Key points

- The full benefits of the HACCP plan will only be realised when it has been properly implemented.
- Implementation is carried out through personnel training, setting up of monitoring systems and completion of support activities.
- Once the HACCP plan is implemented it becomes part of the day to day operations.

The completion and validation of the HACCP plan is often such a relief that it is very tempting to assume that HACCP is complete. However, unless the HACCP system is in place and operating, i.e. the HACCP plan is properly implemented, there will be no real benefit from all the work carried out so far. To make HACCP work in practice it is necessary to ensure that it becomes part of the everyday operating procedures.

'Implementation' is not a HACCP principle. It is not even proprietary to HACCP in that it is an activity requiring many of the normal management practices associated with implementing any system into a business.

Time and money are usually the main restricting factors in every organisation whenever a new system has to be implemented and HACCP is no exception. Costs of implementation should therefore be considered in addition to those of the HACCP study itself (see also Section 1, Frequently asked questions). Sufficient resources must be made available to ensure that the CCPs are effectively implemented and monitoring records are kept.

The implementation process at key stage 3 of HACCP application is best achieved by breaking it down into key steps as shown in *Fig. 16*. Responsibility for implementation of the HACCP plan should be allocated to relevant personnel by the management team in discussion with the HACCP team. Relevant disciplines must be included to deal with training, production, engineering and technical issues. A timetable should then be put into place to organise and carry out training and confirm that monitoring systems, facilities and equipment are in place.

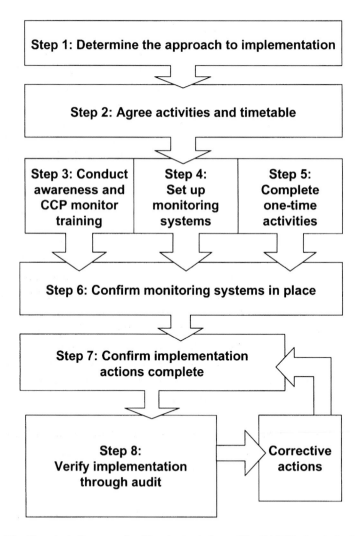

Fig. 16 Key stage 3: example of implementation of the HACCP plan (adapted from Mortimore and Wallace 1998).

Step 1: Determine the approach to implementation

There are two main approaches to implementing the HACCP system and these are either doing it all at once or in phases. The first method involves implementing everything in one go on a certain date, whereas the second allows each section (e.g. a HACCP plan module or even a single CCP) to be implemented independently when the previous section has been completed. There are advantages and disadvantages in both methods but the phased method is likely to be more practical for most businesses.

Step 2: Agree activity list and timetable

As implementation will require the involvement of a large number of people and may take some time to complete, it is useful to construct a detailed activity list and timetable. This should include details of each activity, who is involved and/or responsible for making it happen as well as the deadline for completion. The activity list and timetable can be constructed in the form of a Gantt chart (*Fig. 13*) and since some activities may require others to have been completed, a dependency listing such as a PERT chart may be helpful.

The main activities are likely to include the training and setting-up of monitoring systems. However other activities may have been identified during the HACCP study, which need to be completed in support of the HACCP plan. These will often be issues which came through the decision tree loop at Q1a (*Fig. 15*) and require a modification to the process or product, or the development of a new procedure. It may be that these modifications are directly involved with the identified CCPs or they may have caused hazards to drop out of the decision tree at Q3 by designing them out of the process. These modifications may be considered as 'one-time activities' because they need to be done once either to build in control systems or to design out hazards. Examples of one-time activities include engineering work on plant and equipment.

A requirement to strengthen prerequisite programmes may also have been identified as part of HACCP development and issues highlighted should also be included on the activity list if not already completed.

Step 3: Conduct awareness and CCP monitor training

It is essential that all personnel involved in CCP monitoring receive the necessary training and, most importantly, understand their role in the running of the system. As working practices may be changing it is beneficial for all personnel within the organisation to have a basic understanding of how HACCP works and how it affects their particular working environment. Revisiting the need for compliance with programmes such as good hygiene practices will help personnel understand how their commitment to such programmes links to HACCP and food safety management.

Training is essential but it does not have to be costly. HACCP team members, after appropriate trainer training, should be able to provide in-house training to other personnel and conduct briefing sessions. Training resource materials such as introductory videos and/or materials developed from this book may be helpful, along with examples from the HACCP plan(s) developed for the operation.

Specific HACCP training will be essential for CCP monitors, their deputies, supervisors and managerial staff. This type of training is most likely to be gained through a combination of classroom and on-the-job teaching. They will need to understand what they are expected to do and also why they are doing it and how it fits in with the rest of the system. They will need to have a clear understanding of the principles of CCP monitoring and the corrective actions required when a deviation occurs, as well as how they are expected to record their results or actions. The CCP monitors should understand exactly what a deviation is, when to report it and to whom. It is helpful if this information is specified on the monitoring log sheets or in separate work instructions. In turn the organisation needs to be sure that the individuals chosen are capable of carrying out such responsibilities.

Step 4: Set up monitoring systems

Setting up of monitoring systems requires the development of monitoring instructions and preparation of relevant equipment and recording datasheets for use by the trained CCP monitors. The validated HACCP plan will already detail the monitoring requirements; now these requirements need to be translated into everyday practical activities that can be carried out during production. It is not necessary to reinvent paperwork and record keeping systems if existing monitoring sheets can be adapted, e.g. by adding additional columns for CCP data and signing off.

Consideration should also be given to additional requirements such as the need for extra facilities, e.g. test areas, log sheet storage, computer work stations and work instruction displays where necessary.

Step 5: Complete one-time activities

This requires the personnel responsible for each activity to complete their individual actions so that they can be checked off from the list. This will often include the completion of a diverse range of activities such as engineering work, procedure writing, prerequisite development and additional training. As this may take some time, it is useful for the HACCP team to review progress on a regular basis.

Step 6: Confirm monitoring systems are in place

The monitoring requirements have already been defined within the HACCP control chart and set up earlier. At this stage it is necessary to confirm that they are in place and can be done at the required frequency defined in the HACCP plan.

Step 7: Confirm implementation actions are complete

Once the training and setting up of monitoring systems is confirmed and one-time activities have been completed, the HACCP plan can be transferred into everyday practice through:

- Monitoring CCPs.
- Taking the required actions.
- Recording the results.

This is where HACCP can be said to be implemented and management of the CCPs becomes the responsibility of personnel within the day to day operation.

It is a requirement of HACCP that monitoring records are reviewed by a responsible reviewing official. This will most frequently be a supervisor or manager and this is a good opportunity to check that the implementation actions are complete in the early days of the implemented HACCP plan.

Step 8: Verify implementation through audit

Once the system is implemented and a period of records (e.g. 6 months) is available, a verification audit should be carried out. This may be carried out by internal personnel not directly involved with the day to day running of the HACCP plan or by external HACCP consultants.

3.4 Maintenance of the HACCP system

> ● **Key points**
>
> - The effectiveness of a HACCP system in managing food safety is dependent on continuous maintenance.
> - The HACCP plan should be updated and amended at least once annually.
> - Ongoing personnel training is important to ensure that HACCP awareness is maintained.

A HACCP plan will only achieve its purpose in managing food safety if it is kept up to date, i.e. through continuous maintenance. It needs periodic review, updating and amending if it is to remain current and, therefore, effective. Operations change all the time due to factors such as new raw materials, new recipes and products, improved methods, updated equipment and structural changes in the kitchen or factory. New scientific information on hazards may lead to a review of existing controls. It is therefore necessary that the information resulting from such changes is used to update and amend the HACCP plan at least annually.

'HACCP maintenance' is not a HACCP principle but it is important. If the HACCP study is carried out on a product or process that no longer exists then it will be of little value in controlling food safety for the current activities of the business.

Maintenance of the HACCP system can be achieved by following the steps shown in *Fig. 17*, and the activities considered include regular auditing, hazard data analysis, updating and amending the HACCP plan, all of which should be supported by ongoing training and educational requirements.

It is important that refresher training is carried out regularly to ensure that all personnel involved in HACCP system implementation and use are kept aware of changes to the system and the occurrence of new information, particularly with regard to hazards and their control. New staff also need to be trained so that they have the same level of understanding as their colleagues.

CCP monitors and their supervisors need to be trained appropriately fol-

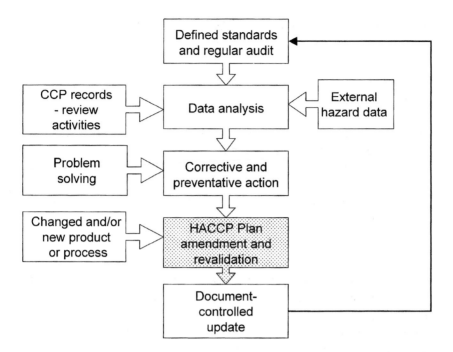

Fig. 17 Keystage 4: Example of verification and maintenance of the HACCP plan (adapted from Mortimore and Wallace 1998).

lowing any amendments to the HACCP plan and it is essential that they understand the reasons behind such changes.

Like verification, HACCP system maintenance is an ongoing activity that will continue through the life of the HACCP system.

3.5 Conclusion

This concludes Section 3. In following all of the steps outlined a business should have a robust HACCP system. It will have been thoroughly researched during its development, be based on sound science and will be kept up to date through proactive maintenance. Food safety is not negotiable. As food business operators, whether in catering, manufacturing, food distribution or enforcement, we owe it to the consumer to do the best we can and to ensure that the food we provide is safe and wholesome. There is no such thing as zero risk but we can manage risk to an acceptable level and it is our responsibility to do this whether we work in a large or small business.

Epilogue

> **Key points**
>
> - HACCP will only succeed if it is backed by management that fully understands the concept.
> - To avoid failure a HACCP system must be planned, executed and implemented correctly. Once implemented, a HACCP system must be reviewed regularly.
> - The people who develop a HACCP system must have the appropriate education, experience and skills.
> - Good communications within the supply chain will result in shared knowledge of potential hazards, better controls and safer food production.

To those of you who have read the entire book before beginning on the Epilogue – have we achieved our objective as set out in the Preface? Have you gained a working knowledge of HACCP? We hope so.

You should also have an understanding (however vague) that the topic of HACCP is not exactly black and white. The HACCP approach is really a way of thinking and working. Learning about HACCP is not like learning about hygiene, which is more fact-based. HACCP is evolving all the time and despite having been around for 40 years it is still the subject of much discussion and debate among professionals, enforcers and academics. In closing the book it seems appropriate to do several things:

- Provide a quick reminder of what we have told you so far.
- Indicate some of the typical problem areas or pitfalls encountered by businesses starting to use HACCP.
- Look at some of the current issues causing debate or controversy.

- Indicate what might come next in terms of businesses wanting to manage food safety.

What we have told you – i.e. HACCP in a nutshell

HACCP is an acronym for the Hazard Analysis and Critical Control Point system. HACCP is recognised by WHO as being the most effective way of preventing foodborne illness. It is a preventative approach to food safety (not quality) management and works by:

- Breaking a process down into individual process steps.
- Analysing each step to find out whether a significant hazard might be introduced – a hazard analysis.
- Deciding what control measures would prevent or eliminate the potential hazards or reduce them to an acceptable level.
- Establishing where it is absolutely *essential* to control the safety of the food, i.e. the CCPs at which the control measures are working.
- Establishing what the critical operating specification is for the critical control points – the critical limits.
- Determining how to monitor the CCPs and what corrective actions are needed in the event of deviation.
- Establishing procedures to verify that the whole system is working, i.e. it will control food safety and is working in practice.
- Documenting all of the above (a HACCP plan) and keeping records.

All of the above are contained within seven HACCP principles as published by Codex (1997b) and NACMCF (1997).

Effective food safety management requires additional skills and activities other than HACCP alone. These can be grouped under the headings:

- Prerequisite hygiene programmes.
- Normal management practices.
- Quality management systems.

A HACCP system must be planned, developed, implemented and maintained by people who have the right skills and a clear understanding of their roles.

Typical problems in using HACCP – i.e. potential pitfalls or why HACCP may fail

Just because a business uses the HACCP principles and has a HACCP plan, it does not mean that they have a safe food process. Why is this?

Lack of real management commitment

Real management commitment is essential if time and money are to be made available to do a thorough job. Much more time is needed to develop and implement HACCP than many managers imagine and they need to know that it takes months rather than weeks in most cases, in order to limit over-expectations. The cost of any training and/or external consultancy and analysis also has to be taken into account.

The attitude and behaviour of employees could be a barrier to the implementation of HACCP. Attitudinal problems can include resistance to change but also resistance to authority in general. This could be one of the causes of behavioural problems. As well as HACCP monitors, the prerequisite hygiene controls, which are an essential foundation, demand a certain behaviour. Yet, operatives with an attitude problem will often only operate the required procedures when a manager or supervisor is standing over them. This can include basic hand-washing and personal hygiene as well as CCP control activities. It needs real management commitment to pro-actively identify these behaviours and deal with them.

The low skill level of the food business operator and HACCP team

A HACCP system will only be as good as the people who developed it and the support it receives. Unless it is properly planned, implemented and maintained it will not serve its purpose of managing food safety and it will only give a false sense of security.

One of the main reasons why a HACCP system may fail to meet its objectives is the team not fully understanding its principles. If HACCP is to be a success the management must wholeheartedly understand the concept and believe in the benefits that it will bring to the business.

It is important that the appropriate educated, experienced and trained people are available when a HACCP system is developed. Education and experience are particularly important in being able to analyse hazards correctly and set critical limits for the control measures. While everyone should be able to understand the concept of HACCP, the application of sound science is needed to be able to use it in practice. Training is important for the understanding of the concept in the first place. Some-times, and probably fairly often, a business will see training as a single intervention, i.e. 'we'll send someone on a course and they'll be our expert'. Very few training organisations would claim this as an outcome of their introductory level courses. To be able to take on such a complex project as HACCP, an ongoing series of training interventions would provide a better grounding. This will include additional courses, use of experts as HACCP study facilitators, reading books, using the internet to look at generic HACCP plans, attendance at symposia and conferences, and, very importantly, talking to others who have done it before. Lack of management skills is perhaps as big a reason for failure. HACCP development and implementation can be quite a big project in any business regardless of size. Basic administrative and organisational skills can help keep documentation to a minimum and ensure that time and money are spent wisely.

Over-documentation and complexity

This leads on from the above. Failing to keep the system simple and focused on what is really critical can create burdens on any business, regardless of size.

Cost

Financial considerations are sometimes responsible for the failure of HACCP. When a need for carrying out testing or buying equipment is identified it is important that the requirement is considered carefully and all alternatives are evaluated to ensure that resources are available.

When a gap analysis of prerequisite programmes identifies the need for investment it is important to invest on a prioritised basis as a HACCP system needs the support of these programmes to be effective.

Education and training are also expensive whether achieved through the recruitment of more highly qualified staff (where none were previously available), use of educated, experienced consultants or the running of training courses. It all costs money.

Inadequate prerequisite programmes

As previously stated, prerequisite programmes were formerly known as GMP/GHP programmes and many businesses believe that they are in place. However, to support HACCP it is important that they are formal systems that have been planned, developed, monitored and verified in the same way as HACCP itself.

Failure to implement and maintain the HACCP plan

A paper HACCP plan will not be very helpful if left on the office shelf. It has to direct management to implement the identified controls. An implemented HACCP system also needs to be maintained. Lack of maintenance will mean that potential hazards resulting from changes in procedures or equipment are missed.

These pitfalls can only be avoided through good planning, training, use of the required skills and, most of all, management commitment to the HACCP concept.

Current issues and controversies

There are so many of these that this subject could easily form the basis of another book; however, the following provide a feel for some of the current issues.

HACCP and SMEs (small and medium-sized enterprises)

There is considerable debate over whether HACCP can be used by SMEs. The basic issue is actually the skill base. Product safety is *not* negotiable according to the size and location of the business and a food business with just one employee will be perfectly capable of putting in a HACCP system provided that the person is a good manager and is well educated, experienced and trained in prerequisite programmes, HACCP and other quality management systems.

Unfortunately people without this level of education, experience and training run many SMEs.

There are no international food safety standards or regulatory agencies

Currently national governments legislate for food safety and do not always have the same standards. This does cause problems for multinationals and for export/import businesses.

Use of prerequisite programmes

Some people are concerned that use of such an approach might mean that issues that could be CCPs are missed out from HACCP plans. Prerequisite programmes have, actually, been around for longer than HACCP – it is just that the term and the formalisation of their relationship with HACCP is new. This approach should help to make the identification of true CCPs a lot easier.

NACMCF versus Codex Alimentarius

At the most recent updates these two important committees brought their HACCP standards closer together. It is interesting though that the majority of food businesses in the USA are driven by the NACMCF document and in many cases are unaware of the Codex Alimentarius standard that has been adopted by governments in most other countries.

Codex discussion papers

There are a number of Codex discussion papers available at the moment and they serve to raise awareness and open debate in areas which still require development. Two very relevant topics out for discussion at the time of writing are:

- Proposed Draft Guidelines for Evaluating Objectionable Matter in Food (Codex 2000a).
- Proposed Draft Guidelines for the Validation of Food Hygiene Control Measures (Codex 2000b).

What next in terms of food safety management?

Many of these topics are developing rapidly but here are a few indicators.

Risk assessment

At the time of writing Codex has published the draft document on microbiological risk assessment (Codex 1998a). This will almost certainly increase the discussions surrounding this topic. The determination of the acceptable level of risk is a government decision. Risk management is a food business responsibility and risk communication is a joint communication with consumers.

Food safety objectives (FSOs)

This item could also appear under the earlier heading 'Current issues and controversies', as this is also undergoing a great deal of debate. In brief, the idea is that micro-biological FSOs can be used by both food businesses and regulators. They are sometimes expressed as the concentration or prevalence of microbiological levels of pathogens aimed at for foodstuffs. In the case of the food industry, FSOs can be used to establish performance criteria necessary to meet the FSO, and in the case of the regulators they can be used to establish the microbiological criteria that must be met to comply with the FSO for a particular hazard in a product (Mitchell 2000).

FSOs have been used in New Zealand and Australia for a few years now and the approach seems to be gaining acceptance on an international level. The Codex Alimentarius is currently reviewing this topic (Codex 1998b).

The exciting aspect of the FSO debate is that it may lead to agreements on hazard acceptability at government level which could help in removing trade barriers through the principle of equivalency of food safety management systems.

Consumer perceived hazards

Food has never been safer, yet concerns over food safety appear to be increasing. This is largely fuelled by both the media and consumerist associations, yet the industry has to manage these concerns and in doing so has to anticipate consumer perception of food safety and manage that alongside the real food safety issues in order to remain in business.

As we said earlier, there is no such thing as 100% safety; zero risk does not exist. But through tools like HACCP we largely reduce food safety risks to an acceptable level. We may also need to use this type of approach for management of perceived risk.

In addition, consumers are increasingly concerned about ethical issues such as child labour or sustaining economies in certain countries. Once these concerns (perceived hazards, ethical concerns and any other consumer issue) are known it is possible to evaluate them in the same way that we now do for significant food safety hazards, i.e. to analyse the process step by step, seeing where control can be applied and the issue can be prevented from becoming a reality. The same approach as used for HACCP may well extend to manage such issues.

HACCP can be used by all sectors of the supply chain but in reality it has largely been used by the processing industry. We foresee greater acceptance for the need to use an integrated approach, a fully operational matrix of activity across the supply chain (Mortimore & Wallace 1998), with shared hazard analysis between primary producers and their processing industry customers and between them and their retail and foodservice customers, all working towards identifying the areas where hazards may arise in the food chain and, therefore, where the preventative controls can be put in place. Without good communications each segment of the supply chain is ignorant of the matters arising later on.

Education and training

Only just starting to be recognised within government and industry is the need for a review of the effectiveness of food industry education and training. What was recognised as a problem a few years ago (Mortimore & Smith 1998) was the fact that many so called HACCP and food safety trainers were in fact hygiene trainers who had 'jumped on the bandwagon'. Conversely many HACCP experts were good presenters rather than trainers. This is still a problem but has helped to recognise that the approach taken to training within the food industry with regards to food safety assurance is not exactly effective, i.e. a single intervention of food hygiene training (a one-day session) or of HACCP (a two-day course) will not change the behaviour. Much more research is needed in this area.

Enforcement

It is anticipated that as the enforcement officers gradually build up their experience, the effectiveness of their assessment of HACCP systems will be improved. This actually applies not only to regulators' enforcement officers but also to other types of second and third party audit such as retail customers and inspection bodies. Currently some businesses find themselves in the position of having to defend their HACCP plans because the auditors or assessors have less of an understanding of how to do HACCP than does the HACCP team that developed it. This can be very difficult and it is not unknown for companies to have to put in extra 'CCPs' because their customers told them to do so (Mayes & Mortimore, 2001).

These are probably not the only debates, controversies and emerging issues but hopefully by including some of them we have managed to convey the fact that this is an evolving area that offers an outstanding opportunity to be truly efficient, effective and above all practical in the management of food safety across all sectors, technologies and national boundaries. Other issues may include:

- Consumer food safety education.
- Formal third party HACCP certification.
- Chemical FSOs.

What more can we say? We hope that you have enjoyed this book and that your appetite has been whetted for more in-depth discussions and training on this dynamic topic.

Appendix A

Case study: frozen cheesecake

1   Introduction

This case study is provided to illustrate the application of the HACCP principles as discussed in the main text (Sections 2 and 3). It is laid out in the format of a HACCP study, initially giving general information such as HACCP team details, terms of reference and product description, and then progresses through the process flow diagrams, hazard analysis and CCP determination to the hazard control chart.

This case study is a fictional example and is not intended as a generic HACCP plan. It is provided without any liability whatsoever in its application and use.

2   The company

The company is a medium-sized frozen foods company producing a range of desserts. Production is mainly automated but manual processes are used for decorating/finishing of products.

3   HACCP team members

- Technical/QA manager
- Production manager
- Line supervisor
- QC supervisor
- Maintenance manager

4   Terms of reference

- The HACCP study covers all types of food safety hazards: biological, chemical and physical.
- The HACCP system is supported by prerequisite good manufacturing practices throughout the factory. This includes a high care area for post-baking activities where different colour clothing is worn and tight hygiene controls are practised.
- This study covers a range of frozen cheesecakes.
- In this study the entire process is divided into six modules.

5   Product description

General

- Frozen ready-to-eat products to be consumed after thawing at ambient temperature for 4 hours or in a refrigerator overnight.
- After thawing the product must be kept chilled and consumed within 24 hours, so temperature and/or time abuse is potentially high.
- The product is targeted at the general public and may be consumed by high risk individuals, e.g. children and elderly people.
- The range comprises the following flavours: vanilla, chocolate, chocolate and hazelnut, strawberry and blackcurrant.

Raw materials

The raw materials used are as follows.

Chilled

- Dairy ingredients: soft cheese in plastic buckets; cream in stainless steel, mobile, returnable tanks.
- Egg (whole, liquid, pasteurised, sugared) in blue-coloured plastic bags inside buckets.
- Margarine (sunflower-oil based) in blue plastic-lined cardboard boxes.

Ambient

- Chocolate (chips and flakes) and biscuit crumb in blue plastic-lined cardboard boxes, delivered ambient but stored chilled to protect quality and prevent temperature rise of product during production.

- Chopped hazelnuts in plastic-lined cardboard boxes, stored in separate area to minimise the risk of cross-contamination.
- Dry powders, sugar, flour and starch, in paper sacks.
- Fruit toppings, in plastic buckets, aseptically packed, low pH.
- Various liquid flavourings, in plastic containers.

Intrinsic factors

- pH of cheese layer = <5.5.
- Water activity (α_w) of cheesecake layer = >0.90.
- No chemical preservatives are used.

This is a frozen product which does not rely on intrinsic factors for stability. It is not intended to be ambient stable or to be stored refrigerated for long periods after defrosting and will carry instructions to store chilled and consume within 24 hours of defrosting.

Key processes

- Mixing – automated and manual.
- Assembly – automated.
- Baking – in double-entry rack ovens.
- Cooling – on racks in blast chiller.
- Decorating – by hand.
- Batch freezing – on racks.
- Packing – automated and manual.

Main hazards to be considered

- Pathogens in raw materials added after baking.
- Survival and growth of spore formers.
- Allergen control (hazelnuts).
- Hazardous foreign material, e.g. metal.

Main control measures

- Supplier control and raw material certification.
- Temperature control (cooking and chilling).
- Cross-contamination prevention.
- Sieving and metal detection.

HACCP

6 Process flow diagram

The overall process is shown on *Fig. 18* (modular system structure) and divided into six modules which are shown in detail in *Figs 19–24* as follows.

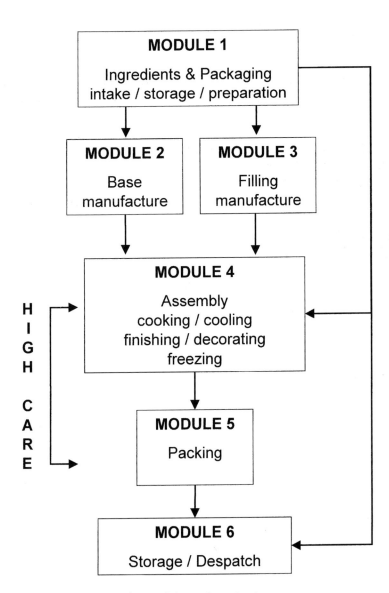

Fig. 18 Frozen cheesecake: modular system structure.

HACCP

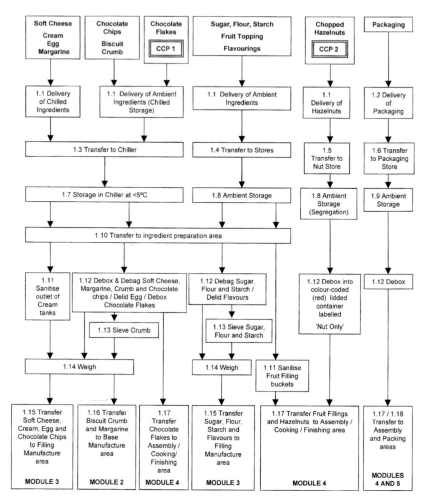

Fig. 19 Module 1: ingredients & packaging intake/storage/preparation.

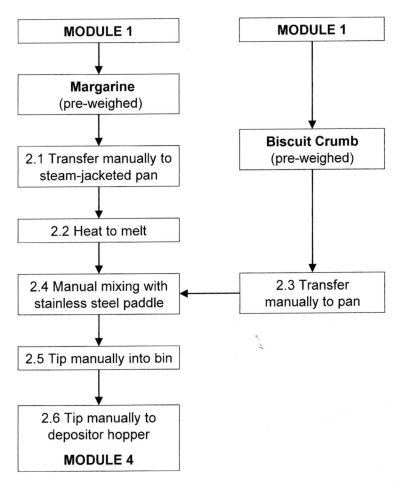

Fig. 20 Module 2: base manufacture.

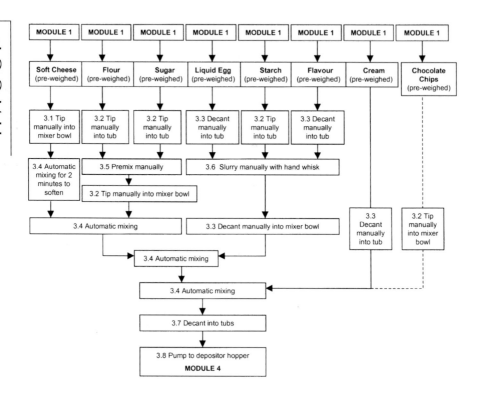

Fig. 21 Module 3: filling manufacture.

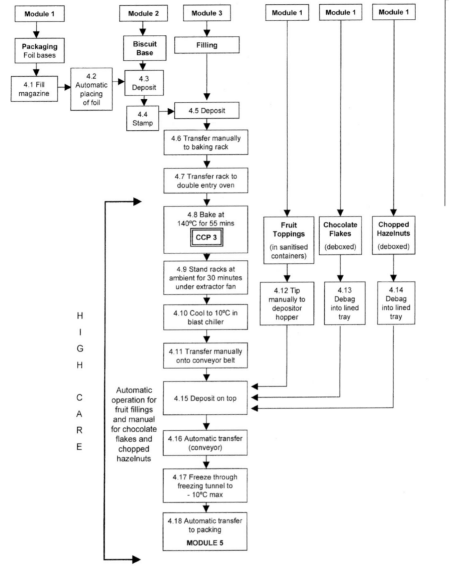

Fig. 22   Module 4: assembly, cooking/cooling, finishing/decorating, freezing.

HACCP

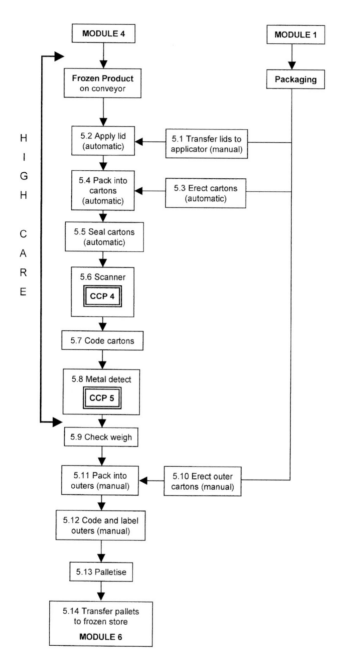

Fig. 23 Module 5: Packing.

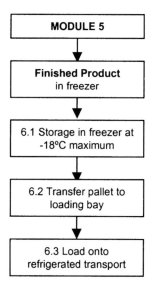

Fig. 24   Module 6: Storage and despatch.

7 Hazard analysis and CCP identification

Table 8 Cheesecake manufacture: hazard analysis and CCP identification for raw materials and process steps.

Raw material	Hazard	Control measures	Significant hazard?	Q1	Q1a	Q2	Q3	Q4	CCP	Justification
Low fat soft cheese (pasteurised)	Vegetative pathogens, e.g. Salmonella, Listeria	Approved supplier, agreed specification, pH 4.4, effective cooking	No						No	Supplier audited on a regular basis. Product will be cooked.
Double cream (pasteurised)	Vegetative pathogens, e.g. Salmonella, Listeria	Approved supplier, agreed specification, effective cooking	No						No	Supplier audited on a regular basis. Vegetative pathogens destroyed by cooking.
Whole egg (liquid pasteurised, contains sugar 10%)	Salmonella	Approved supplier, agreed specification, effective cooking	No						No	Supplier audited on a regular basis. Vegetative pathogens destroyed by cooking.
Chocolate chips	Salmonella	Approved supplier, agreed specification, effective cooking	No						No	Supplier audited on a regular basis. Product will be cooked.
Chocolate flakes	Salmonella	Approved supplier, agreed specification	Yes	Yes		No	Yes	No	Yes	Although the chance of Salmonella contamination in chocolate flakes is considered to be low, the team was aware of historical evidence of Salmonella in chocolate and therefore decided that it was a significant hazard since the flakes are added after cooking.

Material	Hazard	Control measures	Q1	Q1a	Q2	Q3	Q4	Comments
Chopped hazelnuts	Aflatoxin	Approved supplier, agreed specification, certificate of analysis	Yes	No	Yes	No	Yes	This is considered a significant hazard by the team. Effective supplier assurance is essential to assure control by the supplier.
	Allergens from other nuts at supplier's premises	Supplier control, supplier audit	Yes	No	Yes	No	Yes	Controlled by supplier.
	Shell fragments	Supplier control, supplier audit, visual check on debagging and depositing	No					Not a significant hazard.
Biscuit crumb	No hazard identified		No					Supplier control, supplier audit, visual check on debagging, sieving. (Sunflower-based fat.)
Margarine	No hazard identified		No					High fat content, salt. Material does not support growth of pathogens.
Sugar	No hazard identified		No					Supplier control, supplier audit, visual check on debagging, sieving.
Flour	Vomitoxin	Supplier control, supplier audit, certificate of analysis	No					Supplier control, supplier audit and certification, visual check on debagging, sieving.

Q1. Do control measures exist? Q1a. Is control at this step necessary for safety?
Q2. Is the step specifically designed to eliminate or reduce the likely occurrence of a hazard to an unacceptable level?
Q3. Could contamination with identified hazard(s) occur in excess of acceptable level(s) or could these increase to unacceptable levels?
Q4. Will a subsequent step eliminate identified hazards or reduce the likely occurrence of a hazard to acceptable level(s)?

APPENDIX A

Contd

Table 8 *Contd.*

Raw materials	Hazard	Control measures	Significant hazard?	Q1	Q1a	Q2	Q3	Q4	CCP	Justification
Starch	No hazard identified								No	Supplier control, supplier audit, visual check on debagging, sieving.
Flavours	No hazard identified								No	Material does not support growth of pathogens.
Fruit toppings	Foreign material, stalks, leaves, stones/pips	Supplier control, supplier audit, visual check on debagging.	No						No	
Packaging										
Foil bases	No hazard identified								No	Food grade material.
Waxed cardboard lids	No hazard identified								No	Food grade material.
Boxes, outers	No hazards identified								No	No contact with product.

Process step	Hazard	Control measures	Significant hazard?	Q1	Q1a	Q2	Q3	Q4	CCP	Justification
1.1 Intake of ingredients	Microbiological growth (chilled ingredients)	Temperature control	Yes	Yes		No	Yes	Yes	No	Temperature of delivery must be <5°C.
	Physical contamination	Intact packaging	No						No	Visual check.
	Chemical contamination from exhaust fumes	Vehicle engines switched off during intake	No						No	Controlled by site rules – prerequisite programme.
1.2 Intake of packaging	No hazard identified								No	
1.3 Transfer to chiller	Microbiological growth	Time/temperature control	Yes	Yes		No	Yes	Yes	No	
1.4 Transfer to ambient store	No hazard identified								No	Food hygiene practices are observed – prerequisite programme.
1.5 Transfer to nut storage area	Cross-contamination to other materials	Intact packaging	No						No	Food hygiene practices are observed – prerequisite programme. Nut packaging intact.

Q1. Do control measures exist? Q1a. Is control at this step necessary for safety?
Q2. Is the step specifically designed to eliminate or reduce the likely occurrence of a hazard to an unacceptable level?
Q3. Could contamination with identified hazard(s) occur in excess of acceptable level(s) or could these increase to unacceptable levels?
Q4. Will a subsequent step eliminate identified hazards or reduce the likely occurrence of a hazard to acceptable level(s)?

APPENDIX A

Contd

Table 8 *Contd.*

Process step	Hazard	Control measures	Significant hazard?	Q1	Q1a	Q2	Q3	Q4	CCP	Justification
1.6 Transfer to packaging store	No hazard identified								No	Food hygiene practices are observed – prerequisite programme.
1.7 Chilled storage	Microbiological growth	Temperature control. Use within shelf life	Yes	Yes		No	Yes	Yes	No	Prerequisite programme.
1.8 Ambient storage	Physical contamination Microbiological growth	Clean, dry store. Use within shelf life	No						No	No contamination risk – GMP and pest control in place.
1.9 Storage of packaging	No hazard identified								No	
1.10 Transfer to ingredient preparation area	Physical contamination	Covered containers	No						No	Food hygiene practices are observed – prerequisite programme.
	Microbiological growth within chilled ingredients	Time/temperature control	Yes	Yes		No	Yes	Yes	No	Product will be cooked, storage time insufficient for toxin formation.
1.11 Sanitising of creamtank outlets and fruit filling containers	Microbiological contamination if not properly controlled	Validated effective cleaning programme	No						No	Prerequisite programme

Process step	Hazard	Control measure	Q1	Q1a	Q2	Q3	Q4	CCP?	Comments
1.12 Debox/ debag/delid	No hazard identified							No	Food hygiene practices are observed – prerequisite programme.
1.13 Sieving	No hazard identified							No	Food hygiene and equipment maintenance in place – prerequisite programmes.
1.14 Weighing	No hazard identified							No	Food hygiene practices are observed – prerequisite programme.
1.15 Transfer to filling manufacture area	Microbiological growth within chilled ingredients	Time/temperature control	Yes	Yes	No	Yes	Yes	No	Ingredients will be cooked. Storage time insufficient for toxin formation.
1.16 Transfer to base manufacture area	No hazard identified							No	
1.17 Transfer to finishing (high care) area	Microbiological contamination to other ingredients.	Debox in low risk. Spray containers with sanitiser	No					No	Food hygiene practices are observed – prerequisite programme.
	Nut contamination to other ingredients/ process	Use lidded 'nut only' colour-coded (red plastic) containers. At this stage nuts are still in sealed bags	No					No	

Q1. Do control measures exist? Q1a. Is control at this step necessary for safety?
Q2. Is the step specifically designed to eliminate or reduce the likely occurrence of a hazard to an unacceptable level?
Q3. Could contamination with identified hazard(s) occur in excess of acceptable level(s) or could these increase to unacceptable levels?
Q4. Will a subsequent step eliminate identified hazards or reduce the likely occurrence of a hazard to acceptable level(s)?

Contd

APPENDIX A

Table 8 *Contd.*

Process step	Hazard	Control measures	Significant hazard?	Q1	Q1a	Q2	Q3	Q4	CCP	Justification
1.18 Transfer of packaging to packing area	No hazard identified								No	
2.1 Manual transfer of margarine to pan	No hazard identified								No	Food hygiene practices are observed – prerequisite programme.
2.2 Heat margarine to melt	No hazard identified								No	Food hygiene practices are observed – prerequisite programme.
2.3 Manual transfer to biscuit crumb to mixer bowl.	No hazard identified								No	Food hygiene practices are observed – prerequisite programme.
2.4 Manual mixing of biscuit base using stainless steel paddle	No hazard identified								No	Food hygiene practices are observed – prerequisite programme.
2.5 Tipping of biscuit base into bin, transfer to depositor hopper	No hazard identified								No	Food hygiene practices are observed – prerequisite programme.

Process step	Hazard	Control measures	Q1	Q1a	Q2	Q3	Q4	Comments
2.6 Transfer to depositor hopper	No hazard identified		No					Food hygiene practices are observed – prerequisite programme.
3.1 Manual tipping of soft cheese to mixing bowl	No hazard identified		No					Food hygiene practices are observed – prerequisite programme.
3.2 Manual tipping of solid ingredients	No hazard identified		No					Food hygiene practices are observed.
3.3 Manual decanting of liquid ingredients	No hazard identified		No					Food hygiene practices are observed – prerequisite programme.
3.4 Mixing of filling	Microbiological growth	Time/temperature control. Cooking at later stage	Yes	Yes	No	Yes	Yes	Mixing time not long enough to allow toxin formation.
3.5 Manual premixing of sugar and flour	No hazard identified		No					Food hygiene practices are observed – prerequisite programme.
3.6 Manual slurrying of egg, starch and flavour	No hazard identified		No					Food hygiene practices are observed – prerequisite programme.

Q1. Do control measures exist? Q1a. Is control at this step necessary for safety?
Q2. Is the step specifically designed to eliminate or reduce the likely occurrence of a hazard to an unacceptable level?
Q3. Could contamination with identified hazard(s) occur in excess of acceptable level(s) or could these increase to unacceptable levels?
Q4. Will a subsequent step eliminate identified hazards or reduce the likely occurrence of a hazard to acceptable level(s)?

A P P E N D I X   A

Contd

Table 8 *Contd.*

Process step	Hazard	Control measures	Significant hazard?	Q1	Q1a	Q2	Q3	Q4	CCP	Justification
3.7 Decanting of mixed filling into tubs	No hazard identified								No	Food hygiene practices are observed – prerequisite programme.
3.8 Pumping of filling to depositor hopper	No hazard identified								No	Food hygiene practices are observed – prerequisite programme.
4.1 Transfer of foil bases to magazine	No hazard identified								No	
4.2 Placing of foil bases onto line	No hazard identified								No	
4.3 Depositing of biscuit base	No hazard identified								No	Food hygiene practices are observed – prerequisite programme.
4.4 Blocking of biscuit base	No hazard identified								No	Food hygiene practices are observed – prerequisite programme.
4.5 Depositing of filling	No hazard identified								No	Food hygiene practices are observed – prerequisite programme.
4.6 Transfer to baking racks	No hazard identified								No	Food hygiene practices are observed – prerequisite programme.

Process step	Hazard	Control measure	Q1	Q1a	Q2	Q3	Q4	CCP	Justification
4.7 Transfer of racks to oven	No hazard identified							No	Food hygiene practices are observed – prerequisite programme.
4.8 Baking	Survival of vegetative pathogens	Correct heat process	Yes	Yes	Yes			Yes	No subsequent step to remove hazard.
4.9 Standing of racks under extractor (30 minutes)	No hazard identified							No	Food hygiene practices are observed – prerequisite programme. Time would not allow micro hazard to develop. This step allows initial heat loss and prevents condensation problems in the blast chiller.
4.10 Cooling in blast chiller	Growth of surviving spore formers	Time/temperature control			No			No	Cooling trials demonstrate reduction to 10°C in 90 minutes, therefore spore formation not significant.
4.11 Manual transfer to conveyer belt	No hazard identified							No	Food hygiene practices are observed – prerequisite programme.
4.12 Tipping of fruit filling to depositor hopper	No hazard identified							No	Food hygiene practices are observed – prerequisite programme.

Q1. Do control measures exist? Q1a. Is control at this step necessary for safety?
Q2. Is the step specifically designed to eliminate or reduce the likely occurrence of a hazard to an unacceptable level?
Q3. Could contamination with identified hazard(s) occur in excess of acceptable level(s) or could these increase to unacceptable levels?
Q4. Will a subsequent step eliminate identified hazards or reduce the likely occurrence of a hazard to acceptable level(s)?

APPENDIX A

Contd

Table 8 *Contd.*

Process step	Hazard	Control measures	Significant hazard?	Q1	Q1a	Q2	Q3	Q4	CCP	Justification
4.13 Debagging of chocolate flakes onto tray	No hazard identified								No	Food hygiene practices are observed – prerequisite programme.
4.14 Debagging of chopped hazelnuts onto tray	Contamination to other materials	'Nut only' colour-coded trays used. No debagging while other products exposed	No						No	Allergen control essential. Special 'deep' cleaning procedure to be used for trays. Validated cleaning is part of prerequisite programmes.
4.15 (1) Depositing of fruit filling	No hazard identified								No	Food hygiene practices are observed – prerequisite programme.
4.15 (2) Manual decorating with chocolate flakes	St. aureus transfer from operator's hands	Effective handwashing	No						No	Food hygiene practices are observed – prerequisite programme.
4.15 (3) Manual decorating chopped hazelnuts	Allergen cross-contamination	Product containing nuts packed last. Effective cleaning after packing. Dedicated equipment	No						No	Allergen control essential. Special 'deep' cleaning procedure to be used for the line. Validated cleaning is part of prerequisite programmes.
4.16 Transfer to freezing tunnel	No hazard identified								No	Food hygiene practices are observed – prerequisite programme.
4.17 Freezing	Micro growth	Temperature control	No						No	Growth unlikely in a freezing process.

Process step	Hazard	Q1	Q1a	Q2	Q3	Q4	CCP?	Comments
4.18 Transfer of frozen product to packing	No hazard identified						No	Food hygiene practices are observed – prerequisite programme.
5.1 Transfer of lids to applicator	No hazard identified						No	
5.2 Application of lids	No hazard identified						No	Food hygiene practices are observed – prerequisite programme.
5.3 Erection of cartons	No hazard identified						No	
5.4 Packing into pre-formed cartons	Allergen – hazelnut product packed in wrong carton	Yes		No	Yes	Yes	No	There is a scanner further down the line.
5.5 Sealing of cartons	Subsequent microbiological and physical contamination				No		No	Product is already lidded. GMP used.
5.6 Scanning of sealed cartons	Allergen-containing product placed in wrong container where allergen unlabelled	Yes		Yes	Yes		Yes	Allergen control essential. Scanner will pick up wrong cartons that may be received in stack from printer.

Q1. Do control measures exist? Q1a. Is control at this step necessary for safety?
Q2. Is the step specifically designed to eliminate or reduce the likely occurrence of a hazard to an unacceptable level?
Q3. Could contamination with identified hazard(s) occur in excess of acceptable level(s) or could these increase to unacceptable levels?
Q4. Will a subsequent step eliminate identified hazards or reduce the likely occurrence of a hazard to acceptable level(s)?

APPENDIX A

Contd

Table 8 *Contd.*

Process step	Hazard	Control measures	Significant hazard?	Q1	Q1a	Q2	Q3	Q4	CCP	Justification
5.7 Coding of cartons	Loss of traceability	Correct coding	No						No	Legal control measure.
5.8 Metal detection	Presence of metal not identified	All product passes through a functioning metal detector	Yes	Yes		Yes			Yes	No subsequent step to remove hazard.
5.9 Check weighing	No hazard identified								No	Legal requirement i.e. to meet declared weight and may be managed as a CP.
5.10 Erection of outers	No hazard identified								No	
5.11 Packing into outers	No hazard identified								No	
5.12 Coding and labelling of outers	No hazard identified								No	
5.13 Palletisation	No hazard identified								No	

5.14 Transfer to freezer	No hazard identified	No
6.1 Storage in freezer	No hazard identified	No
6.2 Transfer to loading bay	No hazard identified	No
6.3 Load onto transport	No hazard identified	No

Q1. Do control measures exist? Q1a. Is control at this step necessary for safety?

Q2. Is the step specifically designed to eliminate or reduce the likely occurrence of a hazard to an unacceptable level?

Q3. Could contamination with identified hazard(s) occur in excess of acceptable level(s) or could these increase to unacceptable levels?

Q4. Will a subsequent step eliminate identified hazards or reduce the likely occurrence of a hazard to acceptable level(s)?

APPENDIX A

8 HACCP control chart

Table 9 Cheesecake manufacture: HACCP control chart.

Raw material/ process step	CCP no	Hazard to be controlled	Control measure	Critical limits	Monitoring			Corrective action	
					Procedure	Frequency	Responsibility	Procedure	Responsibility
Chocolate flakes	1	Salmonella	Approved supplier	Buy only from approved supplier	Check approved supplier list	Each delivery	Stores supervisor	Reject delivery	Stores supervisor
			Agreed specification	Absent/25g	Check C of A for evidence of compliance	Each delivery	Stores supervisor	Reject delivery	Stores supervisor
Chopped hazelnuts	2	Contamination with other nut traces	Approved supplier	Buy only from approved supplier	Check approved supplier list	Each delivery	Stores supervisor	Reject delivery	Stores supervisor
		Aflatoxin	Agreed specification (Aflatoxin only)	4 ppb max	Check C of A for evidence of compliance	Each delivery	Stores supervisor	Reject delivery	Stores supervisor

Baking	3	Survival of vegetative pathogens	Correct heat process 140°C for 55mins	Core temperature 72°C minimum	Calibrated oven chart recorder visual check and sign off	Each batch	QC operator	Quarantine batch Inform line manager Continue cooking or re-cook until 72°C is achieved	Production operator
Scanning of packed product	4	Allergen containing product in unlabelled packaging	All product passes through suitable scanner	Scanner functioning at all times	Check with packaging samples	Start-up and half-hourly on hazelnut product runs	Line operator	Re-check product since previous satisfactory check	Line manager
Metal detection	5	Ferrous metal contamination	Effective metal detection and rejection	Absence of all ferrous metal above 2.5mm Correctly calibrated metal detector working continuously	Must reject 2.5mm ferrous test strip when placed at centre of product	Start up, every 60 minutes and end of production	Line operator	Re-check product since previous satisfactory check	Line manager

APPENDIX A

HACCP

9 Implementation and maintenance

Validation of HACCP plan elements was carried out prior to implementation. Heat penetration studies were carried out on the batch oven to ensure that the required product centre temperature would be achieved. In addition, validation studies were carried out on the scanner to ensure that it was capable of operating at the line speed.

A phased implementation plan starting with module 1 and following the process through to module 6 was then carried out on a departmental basis. The maintenance plan includes:

- Monthly HACCP team meetings to discuss:
  1. Verification activities such as deviations at a CCP, corrective actions, consumer complaints, audit results, micro results.
  2. Changes to the system such as new ingredients/varieties/process changes which would result in alterations to the HACCP plan.
- 6-monthly verification (internal) audits.
- Annual third-party verification by expert consultant.
- HACCP plan revalidation on an annual basis.
- The requirement for an annual training plan.

Appendix B

Acronyms and Glossary

Acronyms

ASC	Assured Safe Catering
CCP	Critical control point
FAO	Food and Agriculture Organisation
FDA	Food and Drug Administration (USA)
FMEA	Failure, mode and effect analysis
FSIS	Food Safety Inspection Service (USA)
FSO	Food safety objective
GATT	General Agreement on Tariffs and Trades
GHP	Good hygiene practice
GMP	Good manufacturing practice
HACCP	Hazard Analysis and Critical Control Point System
IFST	Institute of Food Science and Technology (UK)
ILSI	International Life Sciences Institute
IOS	International Organisation for Standardisation
NACMCF	National Advisory Committee on Microbiological Criteria for Foods (USA)
NASA	National Aeronautics and Space Administration (USA)
PERT	Programme evaluation and review technique
SMEs	Small and medium-sized enterprises
SPC	Statistical process control
SQA	Supplier quality assurance (also referred to as vendor assurance)
SSOP	Sanitation standard operating procedure
USDA	United States Department of Agriculture
VA	Vendor assurance (also referred to as supplier quality assurance)
WHO	World Health Organisation
WTO	World Trade Organisation

Glossary

The sources of these definitions are indicated in brackets. Definitions which do not indicate their source are taken from Mortimore and Wallace (1998).

Audit. A systematic and independent examination to determine whether activities and results comply with the documented procedures; also whether these procedures are implemented effectively and are suitable to achieve the objectives.

CCP decision tree. A logical sequence of questions to be asked for each hazard at each process step. The answers to the questions lead the HACCP team to decisions determining which process steps are CCPs.

Contaminant. Any biological or chemical agent, foreign matter, or other substances not intentionally added to food which may compromise food safety or suitability (Codex 1997a).

Contamination. The introduction or occurrence of a contaminant in food or food environment (Codex 1997a).

Control (noun). The state wherein correct procedures are being followed and criteria are being met (Codex 1997b).

Control (verb). To take all necessary actions to ensure and maintain compliance with criteria established in the HACCP plan (Codex 1997b).

Control measure. Any action and activity that can be used to prevent or eliminate a food safety hazard or reduce it to an acceptable level (Codex 1997b).

Corrective action. Any action to be taken when the results of monitoring at the CCP indicate a loss of control (Codex 1997b).

Critical control point (CCP). A step at which control can be applied and is essential to prevent or eliminate a food safety hazard or reduce it to an acceptable level (Codex 1997b).

Critical limit. A criterion that separates acceptability from unacceptability (Codex 1997b).

Flow diagram. A systematic representation of the sequence of steps or operations used in the production or manufacture of a particular food item (Codex 1997b).

Food allergen. A food substance which, in some sensitive individuals, causes an immune response causing bodily reactions resulting in the release of histamine and other substances into the tissue from the body's mast cells in the eyes, skin, respiratory system and intestinal system. Allergic reactions may range from relatively short-lived discomfort to ana-phylactic shock and death (IFST 1998).

Food hygiene. All conditions and measures necessary to ensure the safety and suitability of food at all stages of the food chain (Codex 1997a).

Food poisoning. Illness associated with consumption of food which has been contaminated, particularly with harmful micro-organisms or their toxins (IFST 1998).

Food safety. Assurance that food will not cause harm to the consumer when it is prepared and/or eaten according to its intended use (ILSI 1997).

Gantt chart. A project implementation timetable. The Gantt chart shows at a glance the timing and dependencies of each project phase.

Gap analysis. Assessment of the current situation to identify any missing items, i.e. specific gaps, from the required situation.

Good manufacturing practice (GMP). The combination of manufacturing and quality control procedures aimed at ensuring that products are consistently manufactured to their specifications (IFST 1998).

HACCP (Acronym for Hazard Analysis Critical Control Point). A system which identifies, evaluates and controls hazards which are significant for food safety (Codex 1997b).

HACCP control chart. Matrix or table detailing the control criteria (i.e. critical limits, monitoring procedures and corrective action procedures) for each CCP and preventative measure. Part of the HACCP plan.

HACCP plan. A document prepared in accordance with the principles of HACCP to ensure control of hazards which are significant for food safety of the food chain under consideration (Codex 1997b).

HACCP study. A series of meetings and discussions between HACCP team members in order to put together a HACCP plan.

HACCP system. The result of the implementation of the HACCP plan (ILSI 1999).

HACCP team. The multi-disciplinary group of people who are responsible for developing a HACCP plan. In a small company each person may cover several disciplines.

Hazard. A biological, chemical or physical agent in, or condition of, food with the potential to cause an adverse health effect (Codex 1997b).

Hazard analysis. The process of collecting and evaluating information on hazards and conditions leading to their presence to decide which are significant for food safety and therefore should be addressed in the HACCP plan (Codex 1997b).

Hazard analysis chart. A working document which can be used by the HACCP team when applying HACCP Principle 1, i.e. listing hazards and describing measures for their control.

Ingredients. All materials, including starting materials, processing aids, additives and compounded foods, which are included in the formulation of the product (IFST 1998).

(Microbiological) Food safety objective (FSO). A statement based on risk analysis expressing the level of microbiological hazard in a food that is tolerable in relation to an appropriate level of protection (Codex 1998a).

Monitor (verb). The act of conducting a planned sequence of observations or measurements of control parameters to assess whether a CCP is under control (Codex 1997b).

Operational limits. Control criteria which are more stringent than critical limits, and which can be used to take action and reduce the risk of a deviation.

Prerequisite programmes (1). Practices and conditions needed prior to and during the implementation of HACCP and which are essential for food safety (WHO 1999).

Prerequisite programmes (2). Procedures including good manufacturing practices that address operational conditions, providing the foundation for the HACCP system (NACMCF 1997).

Primary production. Those steps in the food chain up to and including, for example, harvesting, slaughter, milking, fishing (Codex 1997a).

Quality management system. A structured system for the management of quality in all aspects of a company's business.

Raw materials. Any material, ingredient, starting material, semi-prepared or intermediate material, packaging material, etc. used by the manufacturer for the production of a product (IFST 1998).

Risk. A function of the probability of an adverse health effect and the severity of that effect consequential to a hazard(s) in food (Codex 1998a).

Significant hazards. Hazards that are of such a nature that their elimination or reduction to an acceptable level is essential to the production of safe foods (ILSI 1999).

Specification. A document giving a description of material, machinery, equipment, process or product in terms of its required properties or performance. Where quantitative requirements are stated, they are either in terms of limits or in terms of standards with permitted tolerances (IFST 1998).

Step. A point, procedure, operation or stage in the food chain including raw materials, from primary production to final consumption (Codex 1997b).

Supplier quality assurance (SQA). The programme of actions to ensure the quality of the raw material supply. Includes preparation of raw material and procedures to assess supplier competency, e.g. inspections, questionnaires.

Validation. Obtaining evidence that the elements of the HACCP are effective (Codex 1997b).

Verification. The application of methods, procedures, tests and other evaluations in addition to monitoring, to determine compliance with the HACCP plan (Codex 1997b).

Water activity (α_w). A measure of the availability of water for the growth and metabolism of micro-organisms. It is expressed as the ratio of the water vapour pressure of a food or solution to that of pure water at the same temperature (IFST 1999).

References

Codex Committee on Food Hygiene (1993) *Guidelines for the Application of the Hazard Analysis Critical Control Point (HACCP) System*, in Training Considerations for the Application of the HACCP System to Food Processing and Manufacturing, WHO/FNU/FOS/93.3 II, World Health Organisation, Geneva.

Codex Committee on Food Hygiene (1997a) *Recommended International Code of Practice, General Principles of Food Hygiene*, CAC/RCP 1-1969, Rev 3 (1997) in Codex Alimentarius Commission Food Hygiene Basic Texts, Food and Agriculture Organisation of the United Nations, World Health Organisation, Rome.

Codex Committee on Food Hygiene (1997b) *HACCP System and Guidelines for its Application*, Annexe to CAC/RCP 1-1969, Rev 3 in Codex Alimentarius Commission Food Hygiene Basic Texts, Food and Agriculture Organisation of the United Nations, World Health Organisation, Rome.

Codex Committee on Food Hygiene (1998a) *Draft Principles and Guidelines for the Conduct of Microbiological Risk Assessment*, ALINORM 99/113A Appendix II.

Codex Committee on Food Hygiene (1998b) Discussion paper on *Recommendations For The Management Of Microbiological Hazards For Food In International Trade.* CX/FX/98/10.

Codex Committee on Food Hygiene (2000a) Discussion paper on *Proposed Draft Guidelines for Evaluating Objectionable Matter in Food*, CX/FX/00/13.

Codex Committee on Food Hygiene (2000b) Discussion paper on *Proposed Draft Guidelines for the Validation of Food Hygiene Control measures*, CX/FH/00/12.

DoH (1993) Assured Safe Catering: *A Management System for Hazard Analysis*. Department of Health. HMSO, London, UK.

FSIS (1996) Food Safety Inspection Service – United States Department of Agriculture. 9CFR Part 304 Federal Register, Vol. 61, No. 144, Rules and Regulations.

Griffith C. (1994) *Application of HACCP To Food Preparation Practices In Domestic Kitchens*, Food Control, Elsevier, UK.

IFST (1998) *Food and Drink – Good Manufacturing Practice: A Guide to its Responsible Management*, 4th edition, Institute of Food Science and Technology, London.

IFST (1999) *Development and Use of Microbiological Criteria for Foods*, Institute of Food Science and Technology, London.

ILSI (1997) *A Simple Guide to Understanding and Applying the Hazard Analysis*

Critical Control Point Concept, 2nd edition, International Life Sciences Institute, Europe, Monograph Series, ILSI Europe, Brussels.

ILSI (1999) *Validation and Verification of HACCP*, International Life Sciences Institute, Europe, Monograph Series, ILSI Europe, Brussels.

Mayes, T. & Mortimore, S.E. (2001) *Making the Most of HACCP*. Woodhead Publishing, Cambridge, UK.

Mitchell, R. (2000) *Practical Microbiological Risk Analysis*: How To Assess, Manage and Communicate Microbiological Risks in Foods. Chandos Publishing, Oxford, UK.

MLC (1998) *Hazard Analysis Critical Control Point Training Programme*, Meat & Livestock Commission, Snowdon Drive, Winterhill, Milton Keynes, UK.

Mortimore, S.E. (2001) How to Make HACCP Really Work in Practice, *Food Control*, **12**(4), Elsevier, UK.

Mortimore, S.E. & Smith, R.A. (1998) Standardised HACCP Training: Assurance For Food Authorities. *Food Control*, **9**(2–3) April–June, 141–5, Elsevier, UK.

Mortimore, S.E. and Wallace, C.A. (1998) *HACCP: A Practical Approach*, 2nd edition, Aspen Publishers Inc, Gaithersburg, MD.

NACMCF (National Advisory Committee on Microbiological Criteria for Foods) (1992) *Hazard Analysis and Critical Control Point System* (adopted 20 March 1992), *International Journal of Food Microbiology*, **16**(1–23).

NACMCF (National Advisory Committee on Microbiological Criteria for Foods), (1997) *Hazard Analysis and Critical Control Point Principles and Application Guidelines*, Adopted August 14 1997.

Panisello, P.J. & Quantick, P.C. (2000) *HACCP and its Instruments: A Manager's Guide*. Chandos Publishing, Oxford, UK. Published in *Journal of Food Protection*, USA.

Sperber, W.H. (2001) Hazard Identification: From A Quantitative To A Qualitative Approach. *Food Control*, **12**(4), Elsevier, UK.

The European Community Directive EC 93/43 on *The Hygiene of Foodstuffs*.

USDA (1973) United States Department of Agriculture 1973 updated 1979. *Acidified Foods and Low Acid Foods in Hermetically Sealed Containers*. Code of US Federal Regulations. Title 21.

USDA (1995) *Procedures for the Safe and Sanitary Processing and Importing of Fish and Fishery Products*. 21 CFRA Parts 123 and 1240 Federal Register Vol 60 No 242 Rules and Regulations.

Wallace, C.A. & Williams, A. (in press 2001) Prerequisites: A Help or a Hindrance to HACCP? *Food Control*, **12**(4), Elsevier, UK.

WHO (1998) *Guidance on Regulatory Assessment of HACCP*. Report of a joint FAO/WHO consultation on the role of government agencies in assessing HACCP. WHO/FSF/FOS/98.5, Geneva.

WHO (1999) *Strategies for Implementing HACCP in Small and/or Less Developed Businesses*, World Health Organisation, WHO/SDE/FOS/99.7, Geneva.

HACCP Resources

For those who will be going on to further education, training and practising in this field we offer some suggestions for further information.

Further reading

This book is an introductory text and is not meant as an in-depth practical guide to implementing HACCP; there are other books which do that. Below are a number of texts that are available. We have selected a few prominent titles that are in English unless otherwise stated:

HACCP: A Practical Approach – Mortimore & Wallace, 2nd edn (1998), Aspen Publishers Inc, Gaithersburg, MD.
Written by practitioners rather than academics, this book does exactly what it says in providing a practical step by step guide to developing and implementing a HACCP system in a food plant. Highly acclaimed. Also available in French, German and Spanish.

HACCP and its Instruments: A Manager's Guide – Panisello & Quantick (2000), Chandos Publishing, Oxford, UK.
A recent publication, very straightforward and well written. Provides some thought-provoking ideas and therefore (as intended) is more likely to be suitable for HACCP managers rather than everyone on the HACCP team.

HACCP Users Manual – Corlett (1998), Aspen Publishers Inc, Gaithersburg, MD.
A very good text particularly for the American market or those exporting into it. It has heavy emphasis on the meat, poultry and seafood industries' mandatory requirements for HACCP and contains various guidelines plus the NACMCF (1997) HACCP principles and application guidelines themselves.

How to HACCP – Dillon & Griffith, 2nd edn (1996), MD Associates, Grimsby, UK.
A text aimed at small businesses. Contains coloured schematics and caricatures. Would be suitable also for use in large businesses as a supervisory level handbook.

Practical Microbiological Risk Analysis: How To Assess, Manage And Communicate Microbiological Risks In Foods – Mitchell (2000), Chandos Publishing, Oxford, UK.

A practical guide to this emerging topic of debate. A must-read and must-have for HACCP managers or team leaders.

Codex Alimentarius Food Hygiene Basic Texts – 1997, WHO, Rome.
Contains the material used as the primary reference source for this book. The contents consist of the general principles of food hygiene, HACCP system and guidelines for its application, and principles for the establishment and application of microbiological criteria for foods.

An Introduction to the Practice of Microbiological Risk Assessment for Food Industry Applications – Guideline No. 28, Campden & Chorleywood Food Research, UK. Association Group (2000). Developed by a combined UK industry and government working party.

Useful websites

Much information on HACCP is available on the internet. The publishers of this book, Blackwell Science, have kindly set up a page within their own website for the use of readers of this book and it can be found at www.blackwell-science.com

This page includes links to various websites where useful, up to date information can be found; they include leading organisations such as The World Health Organisation, The UK Institute of Food Science and Technology, The Institute of Food Technologists in the USA, The Royal Institute for Public Health, The Food and Drug Administration of the United States, and many more.

HACCP training and consultancy providers

There are a huge number of organisations offering HACCP training and consultancy, although it can be difficult to assess their level of competence. Many are excellent but there are also a number who have never actually worked in a factory or catering establishment. It would be impossible to give a list of providers here as there are so many; however, here are a few pointers in finding a reliable supplier of training:

- They should be reputable – either through being a respected organisation or though the provenance of the actual trainer or consultant.
- For HACCP training, look for their trainer skills. They should be trained *trainers* not just experienced presenters and they must be knowledgeable and experienced in HACCP principles and practice.
- Some countries have registration schemes for training centres and courses. For example, in the UK – Royal Institute of Public Health (RIPH); and in the USA – International HACCP Alliance at Texas A&M University.
- Food research associations and academic institutions can be useful in suggesting a source of training and library reference material.

Training materials

Training materials are increasingly being published for use by companies who have internal trainers, consultants or training providers. It is difficult to provide a comprehensive list but details of materials can be found through the links provided on the HACCP book page within www.blackwell-science.com

Index